Statistics and Computing

T0137301

Statistics and Computing

David R. Lemmon
Joseph L. Schafer

Developing Statistical Software in Fortran 95

 Springer

David R. Lemmon
The Methodology Center
The Pennsylvania State University
204 East Calder Way, Suite 401
University Park, PA 16802
USA

Joseph L. Schafer
Department of Statistics and
The Methodology Center
The Pennsylvania State University
204 East Calder Way, Suite 401
University Park, PA 16802
USA

Series Editors:

J. Chambers
Bell Labs, Lucent Technologies
600 Mountain Avenue
Murray Hill, NJ 07974
USA

W. Eddy
Department of Statistics
Carnegie Mellon University
Pittsburgh, PA 15213
USA

W. Härdle
Institut für Statistik und Ökonometrie
Humboldt-Universität zu Berlin
Spandauer Str. 1
D-10178 Berlin
Germany

S. Sheather
Australian Graduate School
  of Management
University of New South Wales
Sydney, NSW 2052
Australia

L. Tierney
School of Statistics and Actuarial Science
University of Iowa
Iowa City, IA 52242-1419
USA

Library of Congress Cataloging-in-Publication Data
Lemmon, David R.
    Developing statistical software in Fortran 95 / David R. Lemmon, Joseph L. Schafer.
        p. cm. — (Statistics and computing)
    Includes bibliographical references and index.
    ISBN 0-387-23817-4 (alk. paper)
        1. FORTRAN (Computer program language)   2. Statistics—Data processing.   I. Schafer,
J. L. (Joseph L.)   II. Title.   III. Series.
    QA76.5.L453 2005
    005.13'3—dc22   2004061447

ISBN 0-387-23817-4   Printed on acid-free paper.

Printed in the United States of America.   (MVY)

9 8 7 6 5 4 3 2 1   SPIN 10969267

springeronline.com

Preface

This book grew from our own need to write good computer programs. We work in an interdisciplinary research center whose mission is to develop and evaluate statistical methodologies and disseminate them to a broad scientific community. Computing and software development play an increasingly important role both in our research and in dissemination. When faced with the task of actually writing a statistical program, however, we found ourselves without a road map. We needed an overall strategy.

After initial discussions, we drafted a brief document for our colleagues that contained guidelines for programming in Fortran. We also wrote libraries of Fortran procedures that could be shared among applications and a sample program to serve as a template for other projects. These materials proved extremely valuable. Those who followed the template—even if they had little or no prior experience in Fortran—quickly learned to produce code that was easy to maintain and debug.

At the same time, we wrestled with another crucial issue: how to disseminate statistical software to reach a wider audience. Some of our consumers would have no problem using an old-fashioned program that runs from the command line. But others had grown accustomed to graphical applications with windows, menus, buttons, etc. Still others were clamoring for add-ons to statistical programs such as SAS®. Given the diversity of environments in which data are being analyzed today, it seemed impossible to satisfy everyone without implementing our methods many times. As we began to learn about the Component Object Model (COM), however, we soon realized that packaging statistical routines as COM servers would help us to reach more potential users. With COM, we found that computational procedures would have to be written carefully and methodically, but they would only have to be written once, as a single COM server can be called by many different client programs.

As this book took shape, a number of thorny (but in hindsight not too difficult) obstacles arose that needed special attention. One was the issue of how to invoke array-valued properties in a COM object from S-PLUS®. We learned from Insightful Corporation, the producers of S-PLUS, that this feature was available but undocumented. More generally, we struggled with how to best package Fortran arrays as variants. Our solution, which we implemented in our variant_conversion module, was to follow the practice of Excel and store an array as a variant containing an array of variants.

Another major hurdle was figuring out how to invoke a COM server from SAS. Interoperability with SAS was offered through the SAS Component Language but not through ordinary SAS programs. This prompted us to create the tool we call sascomio, a dynamic-link library that allows a SAS user to write COM client within PROC IML.

Finally, our preferred method for creating COM servers from Fortran code, which relied on a Compaq tool called the COM Server Wizard, became unavailable when the product was acquired by Intel in 2003. We suddenly found that the latter part of the book hinged upon a commercial product that had been discontinued! Fortunately, we were able to reimplement the functions of the COM Server Wizard through a Perl script that creates all of the necessary extra code. This new method for creating COM servers, which is described in Chapter 7, is almost completely automatic and requires very little knowledge of COM on the part of the programmer.

We would like to thank Linda Collins, Susan Murphy, and other folks from The Methodology Center at Penn State who encouraged us to publish this material as a book. Along the way, many individuals tested our ideas, provided suggestions, and helped to solve problems small and large. Brian Flaherty participated in the initial discussions as we developed our own unique style of pseudo object-oriented programming. Hwan Chung, Recai Yucel, and Hakan Demirtas tried these strategies and gave feedback. Jerry Maples did an outstanding job in writing the first version of our Perl script.

Portions of Chapter 8 rely on features of commercial software products that are not well documented. We wish to thank Steve Lionel of Intel for promptly answering our technical questions and giving us permission to modify the old Compaq templates for generating COM server code. Bob Rodriguez and Simon Smith at the SAS Institute were friendly and responsive as we worked out strategies for interfacing COM servers with SAS. Insightful's Jim Schimert helped us to learn about S-PLUS's COM interface. Thomas Baier, the author of rcom, gave helpful tips on calling COM servers from R.

Finally, we wish to thank John Kimmel and the staff at Springer for their help and patience as they waited for the completed manuscript. Three anonymous reviewers gave many excellent suggestions that we took to heart. Based upon their advice, we expanded the introductory material on Fortran 95 in Chapters 2 and 3; added Chapter 4 on computational routines and numerical issues; added Chapter 6 on dynamic-link libraries; and removed all but the most necessary jargon and technical descriptions about COM and Windows® from Chapters 7 and 8.

Support for this project was provided by the National Institute on Drug Abuse grant 1-P50-DA10075 and by the National Institute on Mental Health grant 1-R01-MH60213.

David R. Lemmon   University Park, Pennsylvania
Joseph L. Schafer   September 2004

Contents

1
Motivation

1.1 Why We Wrote This Book

Few statisticians consider themselves to be professional computer programmers, but many find it necessary to write code at one time or another. Many excellent books have been written on topics of statistical computing—techniques for inverting and decomposing symmetric, positive-definite matrices; methods for generating random variates; calculating tail areas from normal, t, and F distributions; and so on (e.g., Kennedy and Gentle, 1980; Thisted, 1988). The rapid growth of computationally intensive approaches to solving statistical problems—such as Markov chain Monte Carlo and the jackknife—has begun to define a new field of study known as computational statistics (Gentle, 2002). Graduate courses offered by statistics and biostatistics departments tend to emphasize these topics, but less time is devoted to the nuts and bolts of how to write a good program. The excellent *Numerical Recipes* series by Press, Flannery, Teukolsky, and Vetterling (1992, 1996) contains a wealth of functions and subroutines, but these books say little about the infrastructure needed to turn computational procedures into robust, useful programs. General textbooks that teach how to program in Fortran, C++, or Java® are essential for learning these languages, but they target a broad audience and contain few examples of direct interest to statisticians.

After perusing these resources, statisticians who need to begin programming are unsure how to start. What they lack is a framework or paradigm for developing statistical applications. Most of us, through trial and error, gradually settle into a self-styled routine for generating code. Eventually we finish a program that seems to work well enough for the immediate purpose, but we are never quite satisfied with the finished product. It frequently crashes and we cannot tell why. In hindsight, we realize that unfortunate decisions made early in the development process locked in some major limitations, and afterward the program is very difficult to change.

Fortunately, good programming styles can be taught and learned. The purpose of this book is to help a statistician, methodologist, or statistically oriented researcher to write a good Fortran program the first time. We hope to make the experience of developing Fortran applications as quick, painless, and rewarding as possible.

1.2   For Whom We Are Writing

1.2.1   Those Accustomed to FORTRAN 77

This book may prove helpful to anyone with a broad interest in statistical computing. However, we have written it in a particular way to address the special needs of certain groups. One such group is those accustomed to FORTRAN 77 who do not use the modern features of Fortran 90 and Fortran 95, because these new features are unfamiliar or the benefits of using them seem unclear. Through this book, we want to help researchers to update their Fortran skills quickly. Those who graduate from FORTRAN 77 to Fortran 95 will soon find that programs are simpler to write, easier to debug, and more reliable. (Following a convention that others have used, we will write "FORTRAN" in small capitals when referring to versions before Fortran 90 and "Fortran" for later ones.)

1.2.2   Those Who Want to Create Windows Applications

Another target group is those who write Fortran console applications—i.e., programs that are invoked from a DOS or Unix command line—but are now interested in programming for the Windows environment. Console applications are highly linear; they perform a set of actions in a predefined sequence and then stop. In contrast, Windows applications are event-driven, reacting incrementally to mouse clicks and keystrokes. An event-driven program needs to be robust; it must allow the user to vary the event sequence, providing helpful guidance and preventing illegal operations along the way. At present, many programmers accustomed to writing console applications have no idea how to create routines that operate in the graphical, point-and-click world of Windows. By adopting the pseudo object-oriented style described in this book, it is not difficult to convert a Fortran console program into a Windows application with a graphical user interface (GUI).

If you want to add a GUI to statistical routines written in Fortran, you have two basic options: create the GUI yourself or find someone to do it for you. Either way, it's a good idea to clearly demarcate the source code, keeping the computational engine separate from the GUI. Using modern development tools (e.g., Microsoft® Visual Studio® .NET™), creating an interface with a professional look and feel is not beyond the reach of those with moderate programming skills. The mechanisms by which the GUI communicates with Fortran computational code can be tricky, however, even for experienced programmers, and have not been well-documented. This book provides a step-by-step guide for bringing statistical applications written in Fortran into the Windows environment. We also present our own distinctive strategy for separating the Fortran engine from the GUI. In our approach, the essential aspects of the event-driven programming style are built into the Fortran part, keeping the GUI short and simple.

1.2.3   Those Who Want to Interface Fortran with Other Applications

A third target group is those who want to develop custom statistical routines that can be called from other applications, such as Excel®, SAS, S-PLUS, SPSS®, and MATLAB®. All of these commercial packages allow the user to write macros or functions for performing a sequence of tasks automatically. Some of them even provide one or more ways to call external routines written in Fortran. Because interoperability is often considered to be an advanced topic, however, documentation for these features tends to be sparse or difficult to penetrate.

The Windows environment provides mechanisms by which programs written in different languages can communicate with each other. Computational routines written in Fortran can be packaged as dynamic-link libraries (DLLs) and loaded into other applications as they are running. Over the years, DLLs have evolved and given way to more refined methods of communication among software components. The Component Object Model (COM) standard has extended the concepts of object-oriented software design beyond the boundaries of a single language. Using COM, a programmer can bundle data structures and computational routines together into a single object and make it available to Windows applications running on the same computer. These applications can pass data to the COM server, invoke computational procedures, and retrieve results in a very straightforward manner.

A recent advance in software development is object-oriented software components that run within *virtual machines* on computers. A virtual machine is a computer program that acts as a small computer within a computer and provides a uniform environment within which programs can run. The best-known examples are Sun's Java language and Microsoft's .NET (pronounced "dot-net") initiative. They allow interoperability across computers and networks. In this book we will develop a graphical user interface using the Visual Basic .NET® language. The technology of DLLs, COM, and .NET, still unfamiliar to many, opens up exciting possibilities for statistical researchers to disseminate new methods to a broad audience.

1.3 The Need for Good Programming Practice

1.3.1 Programming for Research Dissemination

Statisticians are in the business of creating new statistical methods. As data analysis and modeling have become more computer-intensive, implementation and software development have become an essential part of the dissemination process.

In the past, one could propose a new procedure, publish an article about it, and believe that interested readers would successfully carry out the method by hand calculator. Those days are gone. Journals in statistics and biostatistics are now filled with articles that feature computationally demanding optimization routines, genetic algorithms, simulation by Markov chain Monte Carlo methods, and so on. Some of these techniques would be applied more widely by researchers if software were available. However, it takes a very long time for new statistical methods to be incorporated into mainstream statistical packages or more specialized commercial products. Statisticians who wish to make an impact are now finding that they need to take a more active role in programming.

1.3.2 Programming Standards

If a new procedure merits publication in a major journal, shouldn't it be implemented by the authors in such a way that others may actually use it? The following story may sound painfully familiar. An interesting journal article appears with a statement like this: "The method was implemented in a Fortran program that is available from the authors upon request." Upon contacting the authors, one receives a large file of uncommented code that, if it successfully compiles, seems to work only for the single data example that appears in the published article. In most cases, the program was never really intended to be shared with others. By the time the article actually appeared in print, the authors had long since put it aside to work on other projects; their memories have faded, and to go back now and clean up the code or create documentation is simply too daunting.

The solution to this problem must come from two directions. On one hand, editorial boards of scientific journals may consider adopting policies about code sharing. Referees and editors may need to examine code as part of the review process and enforce minimal standards (e.g., requiring a reasonable amount of documentation). Many journals have already created stable Web sites where resources—code, data files, unpublished technical reports, etc.—are made available to interested readers; submission of these resources in an acceptable format may be required before the article appears in print. From the other direction, statisticians embarking on computing projects need to write code that is clean enough and sufficiently documented to share with others.

1.3.3 Benefits of Good Programming Style

Through painful experience, we have now concluded that if a computer program is worth writing at all, it is worth writing well from the start. All too often, a programmer will say to himself, "I just want to get this program working as quickly as possible; I'll clean it up and add documentation later." This is self-delusion. The only real window of opportunity to create useful documentation is while the code is being written. The source code itself should be a document. Variables, functions, and methods should be given descriptive names. The greatest beneficiary of well-designed and documented code is the programmer, who will be able to return to it months or years later and understand it once again.

Good programming style is not difficult to learn, especially if good examples are available. In our research center at The Pennsylvania State University, we have created a template for statistical applications in Fortran 95 for newcomers to imitate. Initially there may be resistance. Some will say, "I don't want to do it your way; it takes too much time." But good programming practice invariably saves time. A well-written program is much easier to debug, maintain, and improve upon than a poorly written one. Novices may not understand why we do things in a certain way, but once they experience the satisfaction of creating a successful application they are quickly converted.

1.3.4 Benefits of Uniformity

From the developers' side, all of our statistical programs have a similar look and feel. Uniformity of style produces many benefits. Multiple programmers can work together on a project more easily. Code modules can be shared among applications. Producing the source code is much less cumbersome because the same basic strategies are being used over and over; much of the writing is simply cut-and-paste. Finally, staff turnover is less likely to cause a project to fall apart. Professional software developers use uniform styles; why shouldn't we?

1.4   Why We Use Fortran

1.4.1   History of Fortran

Statisticians program in a variety of languages. Some are acquainted with modern object-oriented styles and have written programs in Java, C++, or Visual Basic. Others are adept at writing scripts in SAS IML, S, GAUSS, MATLAB, or (for those old-timers) APL. When computational efficiency is an issue, some create subroutines in Fortran or C and call them from S-PLUS or R. And many of us still return to the decades-old tradition of writing stand-alone console applications in Fortran that are invoked by a command line from a DOS or Unix prompt.

FORTRAN was among the first—and arguably, the very first—of the high-level programming languages. Originally developed by a team of researchers at IBM in the 1950s, its name was derived from *Formula Translation* because it was seen as a tool for automatically translating mathematical formulas into machine-level instructions. When different dialects of the language began to proliferate, the forerunner of the American National Standards Institute (ANSI) developed the first standardization in 1966. The next major revision, FORTRAN 77, was released in 1978 and quickly became the dominant language for engineering and scientific work. By the mid-1980s, however, competition from newer languages such as C began to expose serious limitations in FORTRAN and drove another effort to modernize, producing a major extension in 1990 and a minor revision in 1995. After the introduction of the 1990 standard, the name of the language was no longer regarded as an abbreviation but as a proper name in its own right.

1.4.2   Fortran's Advantages

Many current Fortran users acquired their basic skills during the 1970s or 1980s and continue to follow a classic programming style that has changed little over the last quarter century. While their younger colleagues may take pleasure in calling them "dinosaurs," these die hards know that there are still many good reasons to program in Fortran. No matter how sophisticated the operating systems may become in the future, there will always be a way to get back to a command prompt. Fortran programmers have at their disposal a huge number of computational subroutines available in books such as *Numerical Recipes* (Press et al., 1992, 1996), in back issues of journals such as *Applied Statistics*, and in commercially licensed libraries such as NAG and IMSL. Moreover, although the Fortran language itself has vastly improved and many excellent new features have been added, the essential features of the early language remain intact; programmers are still able to incorporate or update their legacy code into modern applications with only moderate effort. Even in modern computing environments,

Fortran remains an excellent choice for high-performance applications in engineering, mathematics, and statistics.

1.4.3 Useful New Features

The reader of this book who is accustomed to FORTRAN 77 will be introduced to many attractive features of modern Fortran, including:

- modules, which help to organize large, complex programs;

- derived types, which allow variables to be bundled together and passed as a single argument;

- dynamic allocation, by which the dimensions of arrays can be determined at run-time;

- pointers, which increase the power and flexibility of the language in countless ways;

- many other syntactical niceties, including optional arguments and function overloading.

1.4.4 What this book does not cover

This book is not a comprehensive guide to the Fortran 95 language. For example, we do not provide a complete listing of all Fortran statements, intrinsic procedures, or input/output format descriptors. Many excellent and complete references on the Fortran language are already available; one that we highly recommend is the text by Metcalf and Reid (1999). Moreover, reference material on the language itself is distributed with many commercial Fortran compilers in printed or electronic form. Most Fortran programmers already have at their fingertips the ability to locate manual pages and help files to quickly answer their specific questions. Rather than writing another Fortran users' manual, we aspire to teach programming strategies and design principles to statisticians who want to create high-quality, reliable software.

1.4.5 Differences Between Fortran 90 and Fortran 95

Fortran 90 was an extensive expansion of FORTRAN 77, but the differences between Fortran 90 and Fortran 95 are rather minor. Nevertheless, there are some good reasons to use Fortran 95 rather than Fortran 90. For example, the `null()` constructor, which was introduced in Fortran 95, is highly useful, as it avoids the tedious initialization of pointers at the start of a program. The example modules and programs included in this book have been tested and validated with Fortran 95, but many of them will not work with Fortran 90. Readers who have Fortran 90 are strongly encouraged to upgrade their compilers; it will definitely be worthwhile.

1.4.6 Pseudo Object-Oriented Programming in Fortran

Unlike C++ and Java, Fortran 95 is not a true object-oriented language. Nevertheless, by adopting certain programming practices, it is possible to mimic many of the essential qualities of the object-oriented style and realize their benefits. The key features of Fortran 95 that make this possible are modules, derived types, and pointers. By adopting the pseudo object-oriented style recommended in this book, the developer gains many advantages of object-oriented programming—for example, the ability to recreate or enhance one part of the program without breaking the functionality of other parts. Adopting this style leads to programs that are more reliable and easier to develop, maintain, and extend.

1.4.7 Fortran 2003

The final committee draft of the next version of Fortran has recently been adopted. The next standard, called Fortran 2003, is a major enhancement of Fortran 95. We allude to some of the features of this new standard, but we do not emphasize them because Fortran 2003 compilers may not be available for some time. None of the language features described in this book have been deleted in the 2003 standard, so our techniques and examples will work for many years to come.

1.4.8 Which Compiler Should I Use?

We do not wish to promote or discourage the use of any specific product. As this book goes to press, we know of six Fortran 95 compilers currently being sold for the Windows environment. We have personally tried three of them—Lahey/Fujitsu™ Fortran (Version 7.1), Salford FTN95™ (Version 4.60), and Intel® Visual Fortran (Version 8.0)—and found them to be excellent. Each of these compilers fully implements the Fortran 95 standard and is suitable for creating console applications and dynamic-link libraries (DLLs). Compilers for Windows that we have not tried are sold by Absoft, The Portland Group, and NA Software.

On the Unix or Linux® side, we have only tried Intel® Fortran. This compiler is suitable for creating console applications and shared objects, which are analogous to Windows DLLs.

For developing COM servers, our experience is limited to Intel (formerly Compaq) Visual Fortran. Our methods and tools for turning Fortran code into a COM server, which we cover at length in Chapter 7, rely on Intel Visual Fortran and Microsoft Visual Studio .NET. Other compilers may be capable of producing COM servers as well, but the procedures may be substantially different.

1.5 Developing Applications for a Broad Audience

1.5.1 Console Applications and COM Servers

It is not hard for statisticians to write programs for their own personal use or for the statistical community; an S-PLUS or R function may suffice. But creating software that proves useful to behavioral scientists, medical researchers, biologists, or engineers can be very challenging. Consumers of statistical methods are accustomed to analyzing data in many different environments. They will say, "Your method sounds interesting, but my colleagues won't try it unless they can do it in SAS." Others are heavily invested in SPSS. Medical researchers may want to store and analyze data from a clinical trial within an Excel spreadsheet. Writing different versions of the same program for these various computing environments is unattractive and often impractical.

Ideally, a statistician would like to implement a method once, packaging it in a form that can be used by a wide audience. What is the best way to accomplish this? Consider a Fortran console application. By adhering to the ANSI standard and avoiding compiler-specific features, the code will be platform-independent and can be compiled and run with little or no modification on Windows, Macintosh, Unix, or Linux machines. (Platform independence is also achievable with Java, but we feel that Java is still a less than ideal tool for statistical computation, and migration to Java is unattractive for those who are accustomed to programming in other languages.) Compiled console applications are still hard to beat in terms of simplicity of development, portability, and speed. Unfortunately, modern computer users have grown so accustomed to working in Windows environments that many will balk at an application with no GUI, and console applications do not readily interact with packages such as Excel or SAS.

In our research center, we encourage our researchers to write Fortran console programs using the template that we have created. The template conforms to the ANSI Fortran 95 standard, so the program can be written, debugged, and tested using whatever facilities the programmer chooses—on a Unix workstation, a Linux computer, or a Windows PC at the office or at home. After the console program is working satisfactorily, we then take one more step: turning the program into a COM server. Creating a COM server makes it possible to call the same computational engine from a wide variety of Windows applications: an Excel spreadsheet; a SAS, SPSS, or S-PLUS session; or a rudimentary or elaborate graphical interface created in Visual Basic at a later time. If the console application is developed from our template, the process of turning it into a COM server is quite simple and can be carried out in a matter of hours using the development tools that we have created in conjunction with the Intel Visual Fortran compiler.

1.5.2   COM Servers and Clients

The Component Object Model (COM) is a programming standard that emerged and evolved during the 1990s. COM began with a technology called object linking and embedding (OLE) to facilitate communication between software products. Users of Windows began to notice that they could put Excel spreadsheets into Word documents, drag and drop text and graphics from one application into another, launch various applications from a Powerpoint presentation, and so on. OLE worked because all of these applications were built from components using a common standard. As OLE evolved into COM, developers of Windows applications outside of Microsoft began to build their own applications on it as well. Another major development was automation, which allows COM objects to be accessed from scripting languages such as Visual Basic for Applications (VBA). VBA is an abbreviated version of the Visual Basic language used by popular data-management programs, including Excel, to write macros.

A COM-compliant software component is called a COM server. A COM server is binary (compiled) code that contains object definitions, called classes, from which objects can be instantiated at run time. COM servers are packaged and distributed to Windows computers as .dll or .exe files. A COM client, on the other hand, is any program, application, or component that uses a COM server's classes. The client may be written in any programming language, as long as both the client and server adhere to the COM standard on a binary level. Because this interface has been standardized, any COM server can potentially be used by any COM client. COM has been widely accepted by professional developers of Windows applications.

Knowledge of COM opens abundant possibilities for statisticians who write programs. Scientific and statistical applications, including SAS, S-PLUS, SPSS, MATLAB, and others, now provide COM client and COM server capabilities. The primary reason why a statistician ought now to consider creating a COM server is that, by developing and maintaining this single component, the same set of computational procedures can be used in an ever-growing number and variety of applications.

In this book, we present an approach to COM server development using Fortran. Although COM itself is rather complicated, we provide guidance, recommendations, and development tools that take care of most of the details.

1.6 Scope of the Rest of This Book

In the remaining chapters, we provide detailed guidance for nonprofessional programmers who want to create quality statistical applications in Fortran. Chapters 2 and 3 review important aspects of the Fortran language, focusing on new features introduced in the Fortran 90 and Fortran 95 standards that can help improve program architecture. These features allow us to write Fortran programs in a pseudo object-oriented style. We illustrate these techniques by developing a generic module for handling run-time errors, which becomes a key component in all the programs that follow.

Chapter 4 describes strategies for writing computational procedures while paying attention to issues of numerical accuracy, reliability, and efficiency. We offer general strategies and practical advice on how to package routines so that they can be easily shared among applications. To illustrate the principles of good design, we show how to create a computational engine for logistic regression.

Chapter 5 is a step-by-step guide for turning computational routines into a reliable, platform-independent Fortran console program. The basic steps involved in a logistic regression analysis—loading the data, selecting model options, running the estimation procedure, examining the parameter estimates, and diagnostic measures—follow a basic pattern that is common to many statistical applications and thus can serve as a template for many projects.

Those who want to go beyond console applications will be interested in Chapters 6–8. Chapter 6 covers the creation and use of Fortran DLLs. DLLs provide a basic mechanism by which a running program—which may or may not be written in Fortran—can communicate with and access the functionality of a Fortran procedure. Interoperability across languages requires careful attention to how data are passed back and forth.

Chapter 7 describes what COM servers are and demonstrates how to create them. This chapter also guides the reader through the steps of turning a working Fortran console program into a COM server.

A COM server is useless without a COM client. Chapter 8 discusses the process of building COM clients for statistical applications. The first client we discuss is a simple Visual Basic for Applications (VBA) script that allows you to perform logistic regression analysis from within an Excel spreadsheet. From there, we present clients in S-PLUS, SAS, SPSS, and MATLAB. Finally we describe by example how to develop a stand-alone Windows application with the GUI written in Visual Basic .NET.

1.7 Our Source Code and Web Site

The strategies described in this book for creating console applications, dynamic-link libraries, and COM servers were developed for internal use within our research center. We have chosen to publish this material as a service to the statistical profession in the hope that others may learn from it and improve upon it. Please feel free to use our ideas and even our source code in your own projects, as long as you acknowledge us in the same way that you would acknowledge any helpful book, journal article, or software product. Our purpose in disseminating these tools is educational, so we cannot assume any liability for their use or misuse. All resources mentioned in this book are available online at our Web site:

`http://methodology.psu.edu/fortranbook`

2

Introduction to Modern Fortran

This chapter and the next provide a crash course in modern Fortran. Some knowledge of programming, such as mild experience with FORTRAN 77, Basic, or C, will be helpful but is not absolutely required. These chapters cover the basic syntax of Fortran and features of the language that are most useful for statistical applications. We do not attempt to provide a comprehensive reference to Fortran 95. For example, we do not list all of the available edit descriptors or intrinsic functions and subroutines. Those details are readily available in reference manuals distributed with Fortran compilers. Rather, we focus on larger concepts and strategies to help the reader quickly build familiarity and fluency.

2.1 Getting Started

2.1.1 A Very Simple Program

A simple Fortran program that generates uniform random numbers is shown below.

```
────────────────────────── uniform1.f90 ──────────────────────────
!######################################################################
program uniform
    ! Generates random numbers uniformly distributed between a and b
    ! Version 1
    implicit none
    integer :: i, n
    real :: a, b, u
```

```
   print "(A)", "Enter the lower and upper bounds:"
   read(*,*) a, b
   print "(A)", "How many random numbers do ya want?"
   read(*,*) n
   print "(A)", "Here they are:"
   do i = 1, n
      call random_number(u)
      print *, a + u*(b-a)
   end do
end program uniform
!################################################################
```

We have displayed the source code within a box to indicate that this example appears within a file maintained on the book's Web site; in this case, the file is named uniform1.f90. The **integer** and **real** statements declare that the variables a, b, and u are to be regarded as floating-point real numbers, whereas i and n are integers. Understanding the differences between these types of variables is crucial. Generally speaking, real variables are used for data storage and computational arithmetic, whereas integer variables are used primarily for counting and for defining the dimensions of data arrays and indexing their elements. Readers with programming experience may already understand the purpose of the **read** and **print** statements and the meaning of the **do** construct, but these will be explained later in this section. We will also explain **random_number**, a new Fortran intrinsic procedure for generating pseudorandom variates.

Style tip ───

Notice the use of **implicit none** in this program. This statement overrides the implicit typing of variables used in many old-fashioned FORTRAN programs. Under implicit typing,

- a variable beginning with any of the letters i, j, ..., n is assumed to be of type **integer** unless explicitly declared otherwise, and

- a variable beginning with any other letter is assumed to be **real** unless explicitly declared otherwise.

Modern Fortran still supports implicit typing, but we strongly discourage its use. With implicit typing, misspellings cause additional variables to be created, leading to programming errors that are difficult to detect. Placing the **implicit none** statement at the beginning forces the programmer to explicitly declare every variable. We will use this statement in all of our programs, subroutines, functions, and modules.

2.1.2 Fixed and Free-Form Source Code

Readers familiar with FORTRAN 77 may notice some differences in the appearance of the source code file. In old-fashioned FORTRAN, source lines could not exceed 72 characters. Program statements could not begin before column 7; column 6 was reserved for continuation symbols; columns 1–5 held statement labels; and variable names could have no more than six characters. These rules, which originated when programs were stored on punch cards, make little sense in today's computing environments.

Modern Fortran compilers still accept source code in the old-fashioned fixed format, but that style is now considered obsolescent. New code should be written in the free-form style introduced in 1990. The major features of the free-form style are:

- program statements may begin in any column and may be up to 132 characters long;

- any text appearing on a line after an exclamation point (!) is regarded as a comment and ignored by the compiler;

- an ampersand (&) appearing as the last nonblank character on a line indicates that the statement will be continued on the next line;

- variable names may have up to 31 characters.

As a matter of taste, most programmers use indentation to improve the readability of their code, but this has no effect on program behavior. Fortran statements and names are not case-sensitive; the sixth line of the uniform generator could be replaced by

```
Integer :: I, N
```

without effect.

By convention, source files whose names end with the suffix *.f, *.for or *.f77 are expected to use fixed format, whereas files named *.f90 or *.f95 are assumed to follow the free format. Programs consisting of multiple source files, some with fixed format and others with free format, are acceptable; however, you are not allowed to combine these two styles within a single file. Some compilers expect all free-format source files to have the .f90 filename extension, even those that use new features introduced in Fortran 95. For this reason, all of the example source files associated with this book have names ending in .f90, even if they contain features of Fortran 95 that are not part of the Fortran 90 standard.

2.1.3 Compiling, Linking and Running

Before a program can be run, the Fortran code must be converted into sequences of simple instructions that can be carried out by the computer's processor. The conversion process is called building. Building an application requires two steps: compiling, in which each of the Fortran source-code files is transformed into machine-level instructions called object code; and linking, in which the multiple object-code files are collected and connected into a complete executable program.

Building can be done at a command prompt, but the details vary from one compiler to another. Some will compile and link with a single command. For example, if you are using Lahey/Fujitsu Fortran in a Windows environment, the program uniform1.f90 can be compiled and linked by typing

```
lfc uniform1.f90
```

at the command line. Once the application is built, the file of executable code is ready to be run. In Windows, this file is typically given the *.exe suffix, whereas in Unix or Linux it may be given the default name a.out. The program is usually invoked by typing the name of the executable file (without the *.exe suffix, if present) at a command prompt.

Many compilers are accompanied by an Integrated Development Environment (IDE), a graphical system that assists the programmer in editing and building programs. A good IDE can be a handy tool for managing large, complex programs and can help with debugging. The IDE will typically include an intelligent editor specially customized for Fortran and providing automatic indentation, detection of syntax errors, and other visual aids such as coloring of words and symbols.

For example, to build the uniform program in the Microsoft Visual Studio .NET 2003 IDE, using Intel Visual Fortran 8.0, begin by creating a new project. Select File → New → Project... from the menu, and the New Project dialog window appears. Specify a Console Application project, and name the project uniform (Figure 2.1). The next window will prompt for further information; select "Application Settings" and choose "Empty Project" as the console application type (Figure 2.2). Next, add the file uniform1.f90 to the project in the "Solution Explorer" under "Source Files." To do this, right-click on the folder "Source Files," choose Add → Add existing item ... from the pop-up menu, and select the source file (or files) to add.

In the "Solution Explorer," double-click on the source file's name to open the file in the code editor (Figure 2.3). To build the program, select Build → Build uniform from the menu, or press the "Build" button, ▦. The IDE will provide output from the build, indicating success or failure. If a compiler error occurred—due to incorrect syntax, for example—the IDE will direct you to the location of the error in the source-code editor.

Once the program has been successfully built, it can be executed from within the IDE by selecting Debug → Start from the menu, by pressing the "Start" button, ▸, or by pressing the F5 keyboard key.

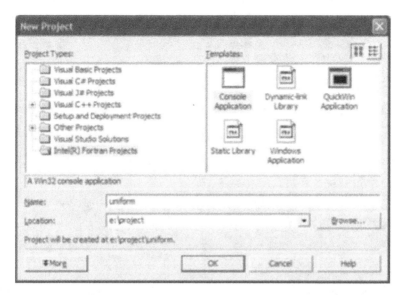

FIGURE 2.1. Intel Fortran New Project dialog.

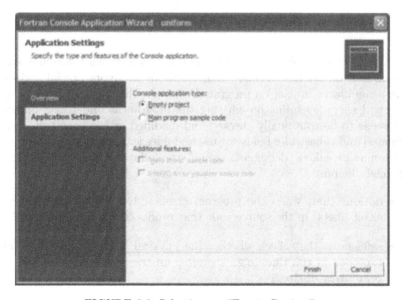

FIGURE 2.2. Selecting an "Empty Project".

2.1.4   Compiler Options

Most implementations of Fortran allow the developer to choose among a
wide variety of options when the code is compiled. Some of these options are
of minor importance and primarily a matter of personal taste—for exam-

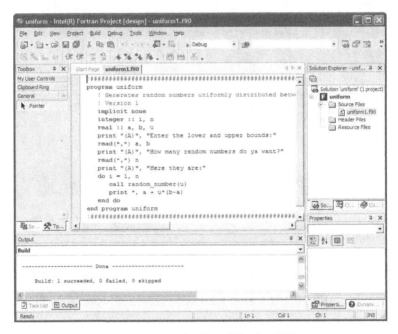

FIGURE 2.3. The Visual Studio IDE.

ple, whether the messages appearing on the screen during the compilation
process will be brief or verbose. Other options are more crucial and can
have a significant impact on program behavior. The optimal choice of options will vary depending on whether the program is under development
and needs to be continually checked and modified or has been tested and
validated and is basically ready for use and release. In the former situation,
which may be called "debugging mode," the following features tend to be
extremely helpful:

- options that, when the program crashes, tell the programmer the
  exact line(s) in the source code that produced the infraction; and

- safeguards that check whether the physical bounds of an array are
  exceeded at run time (e.g., the program tries to access x(4) when x
  is an array of size 3).

An IDE, such as Microsoft Visual Studio, usually provides two possible build configurations: debug and release. The programmer can select
the desired configuration before building and executing the application.
Whenever the debug configuration is used, the IDE can step through your
application's source code one line at a time. It also allows the programmer
to watch the values of the variables (including arrays) at each step. The
programmer can set breakpoints in the editor window at which execution

will automatically pause and then step into subprocedures. These features are extremely helpful for identifying and fixing code errors.

Because these special features embed extra code into a program and cause it to run slowly, they should be turned off when the program is compiled for release. When compiling in release mode, the most useful options are the optimization settings that help to increase execution speed. Optimization features will be discussed in Chapter 4.

2.1.5 Standard Input and Output

Like many simple textbook example programs, **uniform** accepts information from standard input (the computer keyboard) and provides results to standard output (the screen). More elaborate programs that receive and send information via files will appear in Chapter 5. A command-line session that runs the program is displayed below.

```
d:\jls\software\demos>uniform1
Enter the lower and upper bounds:
0 100
How many random numbers do ya want?
5
Here they are:
  3.9208680E-05
  2.548044
 35.25161
 66.69144
 96.30555
```

In this simple program, the statements that perform input/output (I/O) are easy to understand. In the statement

```
print "(A)", "Enter the lower and upper bounds:"
```

"(A)" is a format string indicating that the value of the subsequent expression will be printed verbatim as alphanumeric or character data. In

```
read(*,*) a, b
```

the first asterisk (*) indicates that the data will come via standard input, and the second means that the values for a and b will be read in a list-directed fashion. In list-directed input—also commonly known as "free format" input—multiple items may be separated by any amount of white space (blank spaces, tabs, commas, or new line characters). If one were to apply list-directed input to character variables, like

```
character(len=15) :: first_name, last name
read(*,*) first_name, last_name
```

the user who interacts with the program would have to enclose the string values in single or double quotes:

```
"Bozo" "The Clown"
```

On the other hand, if one specifies alphanumeric input as in

```
read(*,"(A)") first_name
read(*,"(A)") last_name
```

then the user may type

```
Bozo
The Clown
```

without quotation marks, and any leading or trailing spaces on the input lines will be retained in the string variables. String variables will be discussed in greater detail in Section 2.4.

2.1.6 Intrinsic Uniform Generator

Program uniform includes a call to the intrinsic random number generator,

```
call random_number(u)
```

which sets the argument u to a pseudorandom real number x such that $0 \le x < 1$. Every version of Fortran 95 has a uniform random generator that can be called using this syntax, but the details of the implementation vary; compiling and running this program on different platforms could produce different number streams. The argument to random_number may also be a real array, in which case the entire array would be filled with random numbers.

One feature of random_number is that, if the programmer does not explicitly initialize the generator with one or more integer seed values, it will be seeded with processor-dependent integers and produce the same stream of numbers each time the program is run. To produce different results each time, we may include the following line at the beginning of the program, which sets the seeds to values determined by a call to the system clock:

```
call random_seed()
```

This intrinsic subroutine random_seed can also be used to set the seeds to specific values and to query the system to find the current state of the seeds, but the details of this usage may vary slightly from one compiler to another; consult your reference manual for details.

2.1.7 Integer and Real Kinds

In old-fashioned FORTRAN programs, floating-point real variables were declared as real or double precision, the latter being used when round-

TABLE 2.1. Kind parameters for real variables in popular compilers.

	Kind	Bits	Precision (digits)	Comments
Lahey/Fujitsu	4	32	6	default kind
	8	64	15	double precision
	16	128	33	
Salford	1	24	6	default kind
	2	53	15	double precision
	3	64	18	
Intel	4	32	6	default kind
	8	64	15	double precision
	16	128	33	VMS/Unix/Linux only

off error was a potentially serious issue (e.g., when accumulating suffi-cient statistics over large datasets). A **real** variable typically occupied up to 32 bits of storage and was precise to 6 significant digits, whereas a **double precision** variable used up to 64 bits and was precise to at least 15 digits.

The **real** and **double precision** declarations still work, but now the preferred way to determine the precision of real variables is through a **real** statement with the optional **kind** parameter. A variable may now be declared as

```
real(kind=i) :: x
```

or, equivalently, as

```
real(i) :: x
```

where i is a positive integer that determines the overall length of the datum in memory. The available floating-point models and the values of **kind** that correspond to them vary from one compiler to another. The **kind** parameters for some popular compilers are shown in Table 2.1. Omitting the parameter produces a real variable of the default kind.

Integer variables also have **kind** parameters. On most compilers, the default representation is a 32-bit integer that, because the first bit is used for signing, has a maximum value of $2^{31} - 1 = 2{,}147{,}483{,}647$. (In older FORTRAN compilers, the 32-bit representation was often called a "long integer" and had to be specially requested; now it is typically given by default.) The **kind** parameters for integers in some popular compilers are shown in Table 2.2.

To write source code that migrates well from one compiler or platform to another, it is essential to avoid setting the **kind** parameters to specific numeric values such as 3 or 4. This leaves us with two options. The first is to leave **kind** unspecified and pray that the default data types are suf-ficiently large and precise to yield good results. The second, and clearly

TABLE 2.2. Kind parameters for integer variables in popular compilers.

	Kind	Bits	Maximum value	Comments
Lahey/Fujitsu	1	8	127	
	2	16	32,767	
	4	32	2,147,483,647	default kind
	8	64	9,223,372,036,854,775,807	
Salford	1	8	127	
	2	16	32,767	
	3	32	2,147,483,647	default kind
Intel	1	8	127	
	2	16	32,767	
	4	32	2,147,483,647	default kind
	8	64	9,223,372,036,854,775,807	

more desirable, option is to create your own global parameters for desired kinds and use them consistently throughout the source code. For example, if you are using Intel Visual Fortran, you can specify

```
integer, parameter :: my_int = 4
```

and then declare your integers as

```
integer(kind=my_int) :: i, j, k
```

in every procedure. If you want to compile the same code in Salford, you would only need to change the value of my_int from 4 to 3. To make the kind parameters accessible to all parts of the program, it is convenient to keep them in a module; this will be described in Section 3.2. Modern Fortran also has two intrinsic functions, selected_real_kind and selected_int_kind, which can be used to query the compiler and set the kind parameters automatically; use of these functions will be described in Chapter 4.

2.1.8 Do, if, case, goto

Iteration in the uniform program is carried out by a do loop, one of the most frequently used programming constructs:

```
do i = 1, n
   call random_number(u)
   print *, a + u*(b-a)
end do
```

In this do loop, i is the do variable. A Fortran 95 do variable must be an integer. The loop continues to execute as long as the value of the do variable is less than or equal to the upper limit n. By default, the do variable is incremented by one at each cycle, but the increment may be changed by including an optional integer after the upper limit:

```
do i = 2, 2*n, 2    ! gives exactly the same result
   call random_number(u)
   print *, a + u*(b-a)
end do
```

A do loop may be written without a do variable, but the programmer must then provide another way to stop the execution at the desired moment. One way is to include a **while** statement followed by a logical expression:

```
i = 0
do while (i < n)
   i = i + 1
   call random_number(u)
   print *, a + u*(b-a)
end do
```

Another way is to use an if construct with **exit**:

```
i = 0
do
   i = i + 1
   call random_number(u)
   print *, a + u*(b-a)
   if( i == n ) exit
end do
```

The **exit** statement causes the point of execution to jump down to the line immediately following the next **end do** statement; therefore, **exit** within a nested do loop will cause the program to jump out of the innermost loop. A close relative of **exit** is `cycle`, which causes the point of execution to jump up to the do statement above it and begin another cycle of the innermost loop.

 Style tip ──

Notice that in the last two examples, i and n were compared by the relational operators < and == rather than the old FORTRAN operators .lt. and .eq. . The old-fashioned operators still work, but the new ones are preferred. The old and new relational operators are shown in Table 2.3.

───

The syntax of the Fortran if construct is essentially unchanged from previous versions. Conditional execution of a single statement looks like

   if(*logical-expression*) *statement*

whereas conditional execution of a group of statements looks like

TABLE 2.3. Relational operators in Fortran.

Old	New	Meaning
.lt.	<	less than
.le.	<=	less than or equal to
.eq.	==	equal to
.ne.	/=	not equal to
.gt.	>	greater than
.ge.	>=	greater than or equal to

```
if( logical-expression ) then
    statement
    statement
        ⋮
end if
```

or like this:

```
if( logical-expression ) then
    statement
        ⋮
else
    statement
        ⋮
end if
```

Because the if construct depends on a logical expression, it can conditionally execute only one or two groups of program statements. A more flexible structure that handles any number of conditions is the case construct,

```
select case( expression )
    case( value-1 )
        statement
            ⋮
    case( value-2 )
        statement
            ⋮
    case default
        statement
            ⋮
end select
```

where *expression* is a variable or an expression that evaluates to an integer, logical value, or character string, and *value-1*, *value-2*, ... are some or all of the possible values that *expression* may take. The optional case default statement designates actions to be taken if *expression* evaluates to some-

thing other than *value-1*, *value-2*, ... A programming example that uses
the `case` construct will appear in Section 3.6.

Numbered statements have become less common in Fortran, but they
are still useful, particularly in conjunction with `goto`. The use of `goto`
is commonly denounced, with some authors saying that it should never
be used under any circumstances. For the most part, we agree. Frequent
use of `goto` can make a program very difficult to understand and hard to
debug. On the other hand, we find `goto` to be a prudent way to trap errors.
An `if` statement with `goto` can be used to detect an error condition and
immediately jump to the bottom of the procedure to report the error and
exit. The same effect could be achieved without `goto`, of course, but `goto`
enables us to collect all the error-handling statements into one section to
keep our code clean and legible. The use of `goto` for error handling will be
illustrated at the end of Chapter 3. In this book, we use `goto` for trapping
errors but not for any other purpose.

2.1.9 Exercises

1. Write a simple program that converts temperatures from Fahrenheit
   to Celsius and vice versa using $C = 5(F - 32)/9$ and $F = 9C/5 + 32$.

2. Modify one of the uniform generator programs to generate random
   samples from an exponential density

   $$f(y) = \lambda e^{-\lambda y}, \quad y > 0. \tag{2.1}$$

   Use the inverse-cdf method $y = F^{-1}(u)$, $u \sim U(0, 1)$, where F is the
   cumulative distribution function.

3. Write a program that simulates rolling a pair of six-sided dice. It
   should produce 2 with probability $1/6$, 3 with probability $2/36$, and
   so on. Then describe a strategy for generalizing the result to rolling
   n dice, each with k sides.

4. Write a program that simulates random draws from the negative binomial distribution

   $$P(Y = y) = \binom{y-1}{r-1} p^r (1-p)^{y-r},$$

   $y = r, r + 1, r + 2, \ldots$ for any given $p \in (0, 1)$ and positive integer
   r. Use the fact that Y is the number of Bernoulli trials required to
   obtain r successes, with p as the per-trial success rate.

5. Fortran 95 introduced the new intrinsic subroutines `date_and_time`,
   `system_clock`, and `cpu_time`. Learn how to use them and demonstrate each one with a simple programming example.

6. If the birthdays of n individuals are independently and uniformly distributed over 365 days of the year, then the probability that at least two individuals share the same birthday is

$$1 - \prod_{j=1}^{n} \left(\frac{365 - j + 1}{365} \right).$$

Write a program that computes and prints these probabilities for $n = 2, 3, \ldots, 365$.

2.2 Arrays

2.2.1 Rank, Size and Shape

Like its predecessors, modern Fortran supports arrays of logical, integer, real, complex, and character data. A real vector $y = (y_1, y_2, \ldots, y_{10})$ with ten elements may be declared as

```
real :: y
dimension :: y(10)
```

as

```
real, dimension(10) :: y
```

or like this:

```
real :: y(10)
```

Once this array has been created, we refer to the individual elements as y(1), y(2), ..., y(10). The dimension of an array can also be declared as an integer pair i:j, where i is the lower bound (the subscript of the first element) and j is the upper bound (the subscript of the last element). For example,

```
real :: y(0:14)
```

creates an array with elements y(0), y(1), ..., y(14). Even more generally, an array can be dimensioned as i:j:k, where k is the increment between the subscript values of successive elements; for example,

```
real :: y(0:6:2)
```

has elements y(0), y(2), y(4), and y(6).

Multiple dimensions of an array are separated by commas. For example, a matrix with five rows and three columns may be dimensioned as

```
real :: x(5,3)
```

and its elements will then be x(1,1), x(1,2), and so on. Fortran arrays may have up to seven dimensions. The number of dimensions is called the rank. The rank of an array is fixed by the compiler and cannot be changed while a program is running. The Fortran intrinsic function **rank** returns the rank of its argument; if x is declared as shown above, then rank(x) evaluates to the integer 2.

The functions **lbound** and **ubound** return the lower or upper bounds of an array along any or all dimensions. If we dimension an array as

    integer :: iarr(2,0:50)

then lbound(iarr,1) is 1; ubound(iarr,1) is 2; lbound(iarr,2) is 0; ubound(iarr,2) is 50; lbound(iarr) is an integer array with elements 1 and 0; and ubound(iarr) is an integer array with elements 2 and 50.

The size of an array is the number of elements in the entire array or the number of elements along a single dimension. The size of an array can be queried through the intrinsic function **size**. If we declare

    real :: mat(0:5,10)

then size(mat) returns 60, size(mat,1) returns 6, and size(mat,2) returns 10. Taken together, the sizes of all dimensions of an array are called its shape. The intrinsic function **shape** returns the shape of an array as a rank-one array of integers; in this case, shape(mat) is an integer array with elements 6 and 10.

2.2.2 Array Functions

The intrinsic functions **maxval**, **minval**, **sum**, and **product** can find the minimum, maximum, sum, or product of all the elements across a whole array or across any of its dimensions. These functions share the same basic syntax. For example, if x is a 2×3 real array, then:

- sum(x,1) returns a rank-one real array of size 3 whose elements are x(1,1) + x(2,1), x(1,2) + x(2,2) and x(1,3) + x(2,3);

- sum(x,2) returns a rank-one real array of size 2 whose elements are x(1,1) + x(1,2) + x(1,3) and x(2,1) + x(2,2) + x(2,3); and

- sum(x) returns a real scalar equal to the sum of all six elements of x.

These functions can also be applied to integer arrays, in which case the values returned will be of the integer type.

The functions **all**, **any**, and **count** perform similar operations on logical arrays. If x is a logical array, then all(x) returns .true. if every element of x is .true; any(x) returns .true. if at least one of the elements of x is .true.; and count(x) returns an integer value equal to the number of elements of x that are .true.

The generic intrinsic functions `transpose` and `matmul` return whole arrays whose size, type, and kind depend on those of the arguments you provide. If x is a $q \times r$ array, then `transpose(x)` is an $r \times q$ array of the same type and kind. If x is $q \times r$ and y is $r \times s$, then `matmul(x,y)` is $q \times s$. Matrix multiplication usually involves real arrays, but it can also be applied to arrays of integers or logical values, with the result being integer or logical. Before invoking `transpose` or `matmul`, make sure that you have available an array of the correct type and shape to hold the result. The `matmul` function can also be used to multiply a matrix by a vector or a vector by a matrix, but at least one of the arguments must have rank two; multiplication of vectors of the same length is accomplished by `dot_product`, which returns a scalar. Note that `matmul` is designed for arbitrary arguments and may be inefficient for problems with special structure—for example, computing symmetric matrices such as $X^T X$ or multiplying matrices that are known to be diagonal, banded, or triangular.

2.2.3 Operations on Arrays and Array Sections

Modern Fortran can perform some intelligent assignment and comparison operations on entire arrays. If x is an array, then

```
x = 0
```

sets each element of x to zero. If x and y have the same shape, then

```
x = y
```

sets each element of x equal to the corresponding element of y. To check the arrays for conformity, the previous statement could be preceded by

```
if( .not. all( shape(x)==shape(y) ) ) goto 100
```

with subsequent statements to handle the error condition. The result of a comparison operation applied to two arrays of the same shape produces a logical array of that shape. In the previous example, `shape(x)==shape(y)` evaluates to a logical array indicating, for each dimension, whether the sizes of x and y match.

In a similar way, operations can be performed on array sections. The colon (`:`) denotes the entire extent of an array along a particular dimension; for example,

```
sum( x(i,:) * y(:,j) )
```

or

```
dot_product( x(i,:), y(:,j) )
```

computes the inner product of the ith row of x with the jth column of y. A colon preceded or followed by integers limits the range of elements along a dimension. If x is a rank-one array of size n, then

- x(2:4) refers to x(2), x(3), and x(4),

- x(:2) refers to x(1) and x(2), and

- x(1:), x(:n), x(:), and x are equivalent.

Assignment statements that operate on arrays or array sections are convenient for keeping source code concise, but on some systems they may also yield benefits in terms of efficiency because they allow the compiler to choose an order for the elementwise operations to optimize execution. The new Fortran 95 statements **where** and **forall** also signal to the compiler that a group of elementwise operations may be performed in any order for optimal performance. The **forall** statement can replace any do loop in which the order of the operations is unimportant. For example, a sum of squared deviations $\sum_{i=1}^{n} (y_i - a)^2$ can be computed as

```
ssdev = 0.
do i = 1, n
   ssdev = ssdev + ( y(i) - a )**2
end do
```

as

```
ssdev = 0.
forall( i = 1:n ) ssdev + ( y(i) - a )**2
```

or like this:

```
ssdev = sum( ( y - a )**2 )
```

In the last statement, the scalar value in a is implicitly broadcast to an array of the same shape as y prior to the computation of y-a.

The **where** statement carries out an operation on selected array elements corresponding to the .true. values in another array of logical values. For example, if n is an integer array and x is a real array of the same shape, then

```
where( n == 2 ) x = 0.
```

sets each element of x to zero only if the corresponding element of n is 2.

2.2.4   Your Mileage May Vary

Programmers who use array functions such as sum, dot_product, and matmul sometimes find that the resulting performance is slower than if they code these operations themselves using do loops, especially if they pay attention to issues of data storage, loop order, and stride (see Section 4.3). These Fortran intrinsics are guaranteed to work in implementations of the Fortran 95 standard, but they are not guaranteed to be fast. In response, some vendors have begun to offer specially optimized versions of these functions for users who demand high performance. Similarly, the new statements **where** and **forall** were designed to promote efficiency, but depending on your compiler and your computer's architecture, you may not see any improvement over expressions involving whole arrays or array sections with colon notation. You may not even see any improvement over conventional **do** loops because many compilers already have optimization features that can reorganize these loops.

Given these complexities, we do not categorically advise programmers to apply all of these new features in all of their coding. On the other hand, we do not want to discourage their use either because the architects of the Fortran standard had good reasons for adding these features to the language; even when they do not increase efficiency, they may have other important benefits (e.g., keeping source code clean and concise). If you are migrating from FORTRAN 77 to Fortran 95, don't feel obligated to revise your existing code to use these features, as traditional **do** loops still work and are not going away in the foreseeable future.

2.2.5 Array Allocation

One of the major advances since FORTRAN 77 is that the size of an array no longer has to be determined at compilation time; it can now be established while the program is running. This feature, called dynamic allocation, is illustrated by the revised uniform generator program below.

```
———————————————————— uniform2.f90 ————————
!#####################################################################
program uniform
   ! Generates random numbers uniformly distributed between a and b
   ! Version 2
   implicit none
   integer :: i, n
   real :: a, b
   real, allocatable :: x(:)
   print "(A)", "Enter the lower and upper bounds:"
   read(*,*) a, b
   print "(A)", "How many random numbers do ya want?"
   read(*,*) n
   allocate( x(n) )
   call random_number(x)
   x = a + x*(b-a)
   print "(A)", "Here they are:"
   do i = 1, n
      print *, x(i)
   end do
   deallocate(x)    ! not really necessary
end program uniform
!#####################################################################
```

In this program,

```
real, allocatable :: x(:)
```

declares x to be a rank-one array of real numbers whose size has not yet been determined. The statement

```
allocate( x(n) )
```

sets aside a section of the heap (the currently unused portion of memory) large enough to hold n real numbers. The call to random_number then fills the entire array with random variates.

If the allocate statement were removed, this program could crash at the random_number call because the size of x and its location in memory would be undefined. Similarly, applying the intrinsic functions size, shape, lbound, or ubound to an allocatable array that has not yet been allocated may cause a program to crash. To prevent crashes, you can test the array beforehand with the logical function allocated(x), which returns .true. if x has been allocated and .false. otherwise.

Although allocate reserves memory locations, it may not initialize them, so the contents of an array immediately after allocation may not be meaningful. Therefore, after allocating an array, one should not attempt to view or print its contents, use it in an expression, or let it appear on the right-hand side of an assignment statement until the array has been filled with data; doing so could lead to unpredictable results.

The deallocate statement returns memory to the heap. Explicit deallocation of arrays at the end of a program is usually unnecessary because the operating system does that automatically. It tends to be good practice to explicitly deallocate arrays, however, as it forces the programmer to pay greater attention to issues of memory management. To change the size of an allocated array, you need to first deallocate and then reallocate. Applying deallocate to an array that has not yet been allocated will cause a crash, so if the allocation status is in doubt, be sure to check it first with the allocated function.

2.2.6 Exercises

1. Consider the arrays declared below.

   ```
   real :: x(10,4)
   integer, dimension(0:5,0:4,0:3) :: y
   logical :: larr(0:100:4)
   ```

   What are the returned values of rank(y), size(larr), lbound(x,1), ubound(y), and shape(y)?

2. Write a program that draws a random sample of size n without replacement from the population $\{1, 2, \ldots, N\}$ for any $N \geq 1$ and $n \leq N$.

3. Using allocatable arrays, write a program that tabulates and prints the binomial probability mass function

$$f(y) = \binom{n}{y} p^y (1-p)^{n-y}, \quad y = 0, 1, \ldots, n,$$

and the cumulative distribution $F(y) = \sum_{z=0}^{y} f(z)$ for user-specified values of n and p. Use double- precision arithmetic and take reasonable measures to avoid overflows in the computation of $n!$ for large n.

2.3   Basic Procedures

2.3.1   Subroutines

Only very short programs can be effectively written as a single unit. Subroutines became a part of FORTRAN in 1958 and are indispensable for program organization. A re-revised version of our uniform random generator that uses a subroutine is shown below.

```
                              uniform3.f90
!#######################################################################
subroutine fill_with_uniforms(vec_len, vec, lower, upper)
   ! Fills the rank-one array vec with random numbers uniformly
   ! distributed from lower to upper
   implicit none
   ! declare arguments
   integer, intent(in) :: vec_len
   real, intent(in) :: lower, upper
   real, intent(out) :: vec(vec_len)
   ! begin
   call random_number(vec)
   vec = lower + vec * ( upper - lower )
end subroutine fill_with_uniforms
!#######################################################################
program uniform
   ! Generates random numbers uniformly distributed between a and b
   ! Version 3
   implicit none
   integer :: i, n
   real :: a, b
   real, allocatable :: x(:)
   print "(A)", "Enter the lower and upper bounds:"
   read(*,*) a, b
   print "(A)", "How many random numbers do ya want?"
   read(*,*) n
   allocate( x(n) )
   call fill_with_uniforms(n, x, a, b)
   print "(A)", "Here they are:"
```

```
    do i = 1, n
       print *, x(i)
    end do
    deallocate(x)    ! not really necessary
end program uniform
!######################################################################
```

In this example, `vec_length`, `vec`, `lower`, and `upper` are the dummy arguments, whereas `n`, `x`, `a`, and `b` are the actual arguments. Notice that each of the dummy arguments in `fill_with_uniforms` has an `intent` attribute, a highly useful feature introduced in Fortran 90. A dummy argument may be declared to be:

- `intent(in)`, which means that it is considered to be an input and its value may not be changed within the procedure;

- `intent(out)`, which means that it functions only as an output, and any data values contained in the actual argument just before the subroutine is called are effectively wiped out; or

- `intent(inout)`, which means that the argument may serve as both input and output.

If the `intent` attribute is not specified, it is automatically taken to be `inout`. Explicitly declaring the `intent` for each dummy variable is an excellent practice because it forces the programmer to be more careful and instructs the compiler to prevent unwanted side effects. A program will not compile successfully if, for example, an `intent(in)` dummy argument appears on the left-hand side of an assignment statement. We will follow the practice of explicitly declaring the `intent` for all dummy arguments in our subroutines and functions. (The only exception to this rule is pointers, to be discussed in the next chapter; Fortran 95 does not allow `intent` for pointer arguments.)

2.3.2 Assumed-Shape and Optional Arguments

In the previous example, notice that `vec_length`, the size of `vec`, is being passed as an argument. In old-fashioned FORTRAN programs, the dimensions of all array arguments had to be arguments themselves. In modern Fortran, this is no longer the case; we can remove `vec_length` from the argument list and make it a local variable, like this:

```
subroutine fill_with_uniforms(vec, lower, upper)
    ! Fills the rank-one array vec with random numbers
    ! uniformly distributed from lower to upper
    implicit none
    ! declare arguments
```

```
    real, intent(in) :: lower, upper
    real, intent(out) :: vec(:)
    ! begin
    call random_number(vec)
    vec = lower + vec * ( upper - lower )
  end subroutine fill_with_uniforms
```

The dummy argument vec is now called an assumed-shape array; it derives
its shape from the actual argument being passed. If the size of the array
had been needed within the procedure, we could have obtained it with the
intrinsic function size or shape.

Using assumed-shape arrays is an excellent way to shorten argument
lists, reducing the chance that the subroutine will be called incorrectly.
However, if we simply insert this new version of fill_with_uniforms into
uniform3.f90 and remove the actual argument n from the call statement,
the program will not work. The reason is that any function or subroutine
that uses assumed-shape arrays must have an explicit interface. The explicit interface, a concept introduced in Fortran 90, is a new set of rules
governing how actual and dummy arguments interact. Among other things,
it forces the dummy and actual arguments to agree in type, kind, and rank.
The implicit interface used by old-fashioned FORTRAN procedures had minimal checking, which made certain kinds of programming errors difficult to
detect. For this reason, many experienced Fortran programmers now insist
upon using explicit interfaces for all procedures.

It is possible to create an explicit interface for a subroutine by using an
interface block. An easier way is to simply place the subroutine within a
module because any function or subroutine placed in a module is automatically given an explicit interface by Fortran. Modules will be discussed at
length in Section 3.2; for now, we simply illustrate how to place a subroutine within one. Here is yet another version of our uniform generator that
uses a module.

```
───────────── uniform4.f90 ─────────────
!####################################################################
module my_mod
  ! a simple module that contains one subroutine
contains
  !####################################################################
  subroutine fill_with_uniforms(vec, lower, upper)
    ! Fills the rank-one array vec with random numbers uniformly
    ! distributed from lower to upper, which default to 0.0 and
    ! 1.0, respectively
    implicit none
    ! declare arguments
    real, intent(out) :: vec(:)
    real, intent(in), optional :: lower, upper
    ! declare locals
    real :: a, b
    ! begin
```

```
      a = 0.0
      b = 1.0
      if( present(lower) ) a = lower
      if( present(upper) ) b = upper
      call random_number(vec)
      vec = a + vec * (b - a)
   end subroutine fill_with_uniforms
   !###################################################################
end module my_mod
!######################################################################
program uniform
   ! Generates random numbers uniformly distributed between a and b
   ! Version 4
   use my_mod      ! allows the program to use fill_with_uniforms
   implicit none
   integer :: i, n
   real :: a, b
   real, allocatable :: x(:)
   print "(A)", "Enter the lower and upper bounds:"
   read(*,*) a, b
   print "(A)", "How many random numbers do ya want?"
   read(*,*) n
   allocate( x(n) )
   call fill_with_uniforms(x, upper=b, lower=a)
   print "(A)", "Here they are:"
   do i = 1, n
      print *, x(i)
   end do
   deallocate(x)    ! not really necessary
end program uniform
!######################################################################
```

Placing our subroutine in a module also allowed us to apply another new feature, the `optional` attribute, to the dummy arguments `lower` and `upper`. Declaring a dummy argument as `optional` means that the calling program no longer needs to supply an actual argument for it. The intrinsic function `present`, which returns `.true.` if the actual argument has been provided and `.false.` if it has not, is used within the subroutine to define default values for the optional arguments. In this example, if the calling statement were

```
call fill_with_uniforms(x)
```

then x would be filled with random variates uniformly distributed on the unit interval. Notice that in our actual calling statement,

```
call fill_with_uniforms(x, upper=b, lower=a)
```

the order of the actual arguments and dummy arguments differs; the program still works properly, however, because the names of the dummy arguments `lower` and `upper` appear as keywords, enabling the interface to associate the arguments correctly.

Optional arguments and keywords are very helpful when modifying or enhancing subroutines that are already in use by existing programs. It is now possible to add new arguments and options to a subroutine without having to change any of the preexisting `call` statements. These new features require an explicit interface, however, so when using optional arguments or keywords you should make sure that the subroutine is placed within a module.

 Style tip —————————————————————————————

Optional arguments should usually be placed at the end of the argument list, after the required arguments.

Because the names of dummy arguments now function as keywords, it has become more important to choose these names carefully to make them descriptive and meaningful. Well-chosen names will make your source code easier to read and more understandable.

_____

2.3.3   Functions

A function is a procedure that, when it is called, evaluates to a result. For a nontrivial example, let us imagine that the Fortran intrinsic function `sqrt` did not exist. How could we then compute the square root of a real number? The well-known Newton-Raphson procedure solves $f(y) = 0$ by computing

$$y^{(t+1)} = y^{(t)} - f(y^{(t)})/f'(y^{(t)})$$

for $t = 0, 1, 2, \ldots$, where f' denotes the first derivative of f, and $y^{(0)}$ is a starting value. To compute $y = \sqrt{x}$, we can take $f(y) = y^2 - x$, and the iteration becomes

$$y^{(t+1)} = \frac{1}{2}\left(y^{(t)} + \frac{x}{y^{(t)}} \right).$$

This algorithm converges for any $x > 0$, but if $x = 0$ we must be careful to avoid division by zero. Here is a simple function that computes $y = \sqrt{x}$ to single precision using a starting value of $y^{(0)} = 1$:

```fortran
real function square_root(x) result(answer)
    implicit none
    real, intent(in) :: x
    real :: old
    answer = 1.
    do
        old = answer
        answer = ( old + x/old) / 2.
        if( answer == old ) exit
        if( answer == 0. ) exit
```

```
      end do
   end function square_root
```

Here is a very simple program that calls the function:

```
program i_will_root_you
   implicit none
   real :: x, square_root
   read(*,*) x
   print *, square_root(x)
end program i_will_root_you
```

Notice that square_root has to be explicitly declared as real in the main program because it is an external function and because implicit none has been used. If square_root had been placed in a module, Fortran would have given it an explicit interface and this type declaration would have been unnecessary.

Any function can be made into a subroutine by adding one more dummy argument with the attribute intent(out). Similarly, any subroutine with a single intent(out) argument can be reexpressed as a function. Whether one uses a subroutine or a function to perform a task is primarily a matter of taste. In general, we will adopt the following conventions:

- If a procedure cannot fail to produce the desired result, we will express it as a subroutine.

- If a procedure has a possibility of failing for any reason, we will express it as a function that returns an integer 0, indicating successful completion, or a positive integer, indicating failure.

This rule will allow us to develop a unified approach to handling run-time errors, which we will introduce at the end of Chapter 3. In keeping with this rule, let us revise our square-root function to handle the possibility of a negative or zero input value.

```
integer function square_root(x,y) result(answer)
   implicit none
   real, intent(in) :: x
   real, intent(out) :: y
   real :: old
   if( x < 0. ) then
      goto 5
   else if( x == 0. ) then
      y = 0.
   else
      y = 1.
      do
         old = y
```

```
           y = ( old + x/old) / 2.
           if( y == old ) exit
        end do
     end if
     ! normal exit
     answer = 0
     return
     ! error trap
5    answer = 1
  end function square_root
```

The revised calling program is

```
program i_will_root_you
   implicit none
   real :: x, y
   integer :: square_root
   read(*,*) x
   if( square_root(x,y) > 0 ) then
      print "(A)", "I can't handle that."
   else
      print *, y
   end if
end program i_will_root_you
```

2.3.4   Pure, Elemental and Recursive Procedures

A function is called pure if all of its arguments are intent(in). A pure
function cannot modify the values of its actual arguments; results are com-
municated to the calling program only through the function's returned
value. Our first version of square_root was a pure function. We could
have notified the compiler that the function was intended to be pure by
including the keyword pure in our function definition, like this:

```
pure real function square_root(x) result(answer)
```

If we had done this, the compiler would have generated an error message
if the dummy argument x appeared on the left-hand side of an assignment
operation. Any function called within a pure function must also be pure.
Declaring functions as pure helps us to protect ourselves against unwanted
side effects.

   All of the intrinsic functions in Fortran 95 are pure. Many of these functions are also elemental, which means that they can be applied both to
scalars and arrays. When applied to arrays, the result is the same as if the
function had been applied to each element individually. For example, if x
is a real array, then log(x) is a real array of the same shape containing

the logarithms of the elements of x. We can write our own elemental functions by including the `elemental` keyword in the definition. An elemental function must be given an interface, which we can easily do by placing it in a module. For example, if we define our square-root function as

```
module mymod
   contains
   elemental real function square_root(x) result(answer)
      implicit none
      real, intent(in) :: x
      real :: old
      answer = 1.
      do
         old = answer
         answer = ( old + x/old) / 2.
         if( answer == old ) exit
      end do
   end function square_root
end module mymod
```

then we can apply it to a real array of any shape. An elemental function is assumed to be pure.

A procedure is considered to be recursive if it calls itself. We can allow a procedure to call itself by including the keyword `recursive`. For example, here is a recursive function for calculating $x!$ with no argument checking or overflow protection:

```
recursive integer function factorial(x) result(answer)
   implicit none
   integer, intent(in) :: x
   if( x == 0 ) then
      answer = 1
   else
      answer = x * factorial( x-1 )
   end if
end function factorial
```

Recursive procedures can be useful for sorting data, finding sample medians and quantiles, and managing linked lists and other recursive data structures (Section 3.5.3).

2.3.5   On the Behavior of Local Variables

Any variable in a subroutine or function that is not passed as an argument is a local variable, defined only within the scope of that procedure. A local variable can be declared and initialized in the same statement, like this:

```
logical :: converged = .false.
```

This ensures that `converged` is `.false.` when the procedure is invoked the first time. In subsequent calls, however, the initial value of `converged` could vary. The reason is that any local variable initialized in a declaration statement is automatically given the `save` attribute, allowing its value to persist from one call to another. That is, the statement above is equivalent to

```
logical, save :: converged = .false.
```

because `save` is implied. If the procedure is called a second time in the same program, the initial value of `converged` will be the value it had upon completion of the first call. If you want the local variable to be initialized each time the procedure is invoked, you need to include the executable statement

```
converged = .false.
```

within the procedure. On the other hand, if you really do want the value of a variable to persist, it's better to pass the variable as an argument.

Style tip—————————————————————————————
Don't rely on `save` to store persistent data locally within a procedure because it may produce unexpected results if the procedure is embedded in a DLL or in a COM server.

———————————————————————————————————————

Subroutines and functions may also have local allocatable arrays to serve as temporary workspaces. If the shape of the array is not known in advance but is determined inside the procedure, then the array should be explicitly dimensioned within the procedure by an `allocate` statement. It is not absolutely necessary to explicitly destroy the array with `deallocate` because unless the array has been given the `save` attribute, Fortran 95 automatically deallocates the local array when it exits the procedure. However, explicit deallocation is still a good practice.

Alternatively, if the dimensions of a local array enter the procedure as integer arguments, Fortran will allocate the array automatically. This is called an automatic array. For example, if m and n are dummy arguments, then declaring a local array

```
real :: workspace(m,n)
```

will cause `workspace` to be dynamically allocated when the procedure is called. In this case, Fortran will automatically deallocate the array when the procedure is finished.

2.3.6   Exercises

1. Optional arguments to procedures were discussed in Section 2.3.2. A dummy argument that has been declared optional may be passed as an actual argument to another procedure, provided that the corresponding dummy argument in the latter procedure is also optional. Try this in an example.

2. Create a procedure for matrix multiplication using assumed-shape arrays and write a simple program that calls it. Within your procedure, check the dimensions of the argument arrays for conformity.

3. Write a recursive procedure for generating any term of the Fibonacci sequence
$$0, 1, 1, 2, 3, 5, 8, \ldots,$$
in which each element is the sum of the two preceding elements. Then use this procedure in a program that prints out the first n terms for a user-specified n. Does this procedure seem efficient? Explain.

4. Write elemental functions for computing the logistic transformation
$$\mathrm{logit}(p) = \log\left(\frac{p}{1 - p}\right)$$
and its inverse
$$\mathrm{expit}(x) = \frac{e^x}{1 + e^x}$$
for real arrays of arbitrary shape.

5. Suppose that a procedure begins like this:

```
subroutine do_something( arg1, arg2 )
    integer, intent(in) :: arg1
    integer, intent(out) :: arg2
    real, allocatable, save :: x(:)
    allocate( x(arg1) )
```

Explain why this subroutine may cause a program to crash. What can be done to fix it?

6. In a Fortran procedure, the name of a function may be passed as an argument. If we want to pass the name of a real-valued function, for example, the corresponding dummy argument should be declared as a real variable of the appropriate kind. Write a procedure that accepts the name of a function f as an argument and numerically approximates the first derivative,
$$f'(x) \approx \frac{f(x + \delta/2) - f(x - \delta/2)}{\delta},$$
for a given x and $\delta > 0$.

2.4 Manipulating Character Strings

2.4.1 Character Variables

Modern Fortran has many helpful features for storing and manipulating
character data. In statistical applications, these features are useful for handling filenames, reading and processing data from files, interpreting user-
supplied text expressions, handling error messages, and so on.

A character variable can be declared as a single string of a fixed length,

```
character(len=name_length) :: my_name
```

as an array of fixed-length strings,

```
character(len=name_length) :: my_parents_names(2)
```

or as an allocatable array of fixed-length strings:

```
character(len=name_length), allocatable :: &
    all_my_children(:)
```

In these three examples, the integer `name_length` must either be a constant,

```
integer, parameter :: name_length = 80
```

or, if the string or string array appears in a procedure, `name_length` must
be included among the dummy arguments.

Under many circumstances, it is also possible to declare character variables of assumed length (`len=*`). A character-string constant of assumed
length, created as

```
character(len=*), parameter :: &
    program_version = "Beta version"
```

automatically takes the length of the literal string on the right-hand side of
the equal sign. A string that serves as a dummy argument to a procedure
may have an assumed length, in which case it derives its length from that
of the corresponding actual argument.

 Style tip

When writing procedures for character strings, it's an excellent idea to use
dummy arguments of assumed length. This makes the procedures easier to
call and more general, reducing the chance that the procedure will need to
be changed as the program develops.

The colon (:) allows us to access individual characters or substrings
within a character string. Suppose we declare and set a character variable
as follows:

```
character(len=10) :: my_name
```

```
my_name = "Charlie"
```

Then my_name(2:2) is "h", my_name(2:4) is "har", my_name(:4) is "Char", and my_name(4:) is "rlie   ". If strarray is a rank-one string array, then strarray(i)(j:j) extracts the jth character from the ith element.

2.4.2 Assigning, Comparing, and Concatenating Strings

The equal sign may be used for character assignment as follows. If a and b are strings of equal length, then

```
a = b
```

copies the contents of b into a. If a is shorter than b, the initial part of b goes into a and the rest is discarded. If a is longer than b, then the contents of b go into the initial part of a, and the rest of a is padded with blank spaces. Therefore,

```
a = ""
```

has the effect of filling a with space, and

```
a(i:) = ""
```

has the effect of blanking out the ith and all subsequent characters of a.

Strings can also be compared by the relational operators shown in Table 2.3. If two strings are not of the same length, then the shorter one is implicitly padded with blank spaces on the right-hand side before the comparison is made. Two strings are judged to be equal if characters in every pair of corresponding positions agree; for example, ("boy" == "boy ") evaluates to .true., and (a == "") evaluates to .true. if and only if a is entirely blank.

Applying the inequality operators <, <=, >, and >= to character strings is not recommended because the outcome of these operations depends on a collating sequence that may vary from one compiler to another. Instead, we recommend using the intrinsic lexical functions llt (less than), lle (less than or equal to), lgt (greater than), and lge (greater than or equal to) because these functions rely on the universal ASCII collating sequence. Each of these functions takes two character strings as arguments and returns a logical value; for example, the expression lgt(string1, string2) evaluates to .true. if string1 is lexically greater than string2. One character is considered to be greater than another if it appears later in the ASCII sequence; multicharacter strings are compared one character at a time, moving from left to right, with the first nonidentical pair of characters determining which string is the greater one.

Strings are concatenated (i.e., joined together) by the double forward slash (//) operator. For example, the expression

```
"Tom" // " " // "Sawyer" == "Tom Sawyer"
```

evaluates to .true.. . Concatenation is useful for defining literal strings that are too long to comfortably appear in a single line of source code:

```
gettysburg_address = "Four score and seven years ago " &
    // "our fathers brought forth on this continent " &
    // "a new nation..."
```

2.4.3 More String Functions

The Fortran intrinsic function len returns the length of a string, whereas len_trim returns the length without counting trailing spaces. For example, len("boy ") returns the integer 4, but len_trim("boy ") returns 3. The function adjustl returns a character string of the same length as its input, with leading blanks removed and reinserted at the end; adjustr performs the opposite operation, removing trailing blanks and reinserting them at the beginning. Thus adjustl(" boy") evaluates to "boy " and adjustr("boy ") evaluates to " boy".

An extremely useful function is index, which allows us to search for a single character or string within another string. More precisely,

```
index(string, substring)
```

returns an integer indicating the starting position of the first occurrence of substring within string. For example, index("hello","l") evaluates to 3, as does index("hello","llo"). If the substring does not occur within the string, the returned value is 0. This function has an optional third argument, back, whose default value is .false. . Specifying back=.true. will begin the search from the right side rather than the left, returning the starting position of the last occurrence of the substring within the string. The following function, which uses both index and len_trim, removes any existing suffix from a filename and replaces it with a new three-character suffix supplied by the user.

—————————————— suffix.f90 ——————————————

```
!#########################################################################
integer function add_suffix(file_name, suffix) result(answer)
    ! Returns 0 if operation is successful, 1 otherwise.
    implicit none
    character(len=*), intent(inout) :: file_name
    character(len=3), intent(in) :: suffix
    integer :: i
    if( suffix(1:1)=="" ) goto 5              ! suffix begins with blank
    i = index(file_name, ".", back=.true.) ! look for last period
    if( i==0 ) i = len_trim(file_name) + 1 ! if no period was found
    if( len(file_name) < i+3 ) goto 5         ! not enough room for suffix
    file_name(i:) = "." // suffix
    ! normal exit
    answer = 0
    return
    ! error trap
```

```
5  answer = 1
   return
end function add_suffix
!####################################################################
```

2.4.4   Internal Files

Fortran now allows character variables to be used for input and output; data may be written to them or read from them as if they were files. When character variables are used in this manner, they are called internal files. Internal files provide a convenient mechanism to convert numeric data to character representations and vice versa. If str is a character string and a is a variable, then

```
write(str, *) a
```

writes the value of a to str in free format, and

```
read(str, *) a
```

attempts to read a value of a from str. An error will result in the former case if the written value of a exceeds the length of the string, or in the latter case if the contents of str cannot be interpreted as data of the expected type.

For a more detailed example, suppose we are writing a statistical application that operates on a rectangular dataset consisting of p variables. The user may supply character-string names for the variables, but if names are not given, we will create the default names "VAR_1", "VAR_2", and so on. The code below shows how these default names can be created by writing integers to an internal file.

```
integer :: p                   ! the number of variables
integer, parameter :: var_name_length = 8
character(len=var_name_length), allocatable :: var_names(:)
character(len=4) :: sInt
integer :: i

! lines omitted

allocate(var_names, p)
do i = 1, p
   write(sInt, "(I4)", err=5) i
   var_names(i) = "VAR_" // adjustl(sInt)
end do
! error trap
5  continue
```

In the **write** statement, the format string " (I4) " declares that the integer is to be written to a field four characters wide. In the unlikely event that 10,000 or more variables are present, the **err** specifier will cause the point of execution to jump to a trap where the problem can be handled gracefully.

In Fortran **read** and **write** statements, the format string does not need to be a literal character string such as " (I4) "; it may also be a character variable whose value is altered during program execution. This feature greatly facilitates run-time formatting, which was very awkward in FORTRAN 77. For example, in FORTRAN 77 it was difficult to print out a nicely formatted rectangular table for which the number of columns was not known in advance. By writing integers to internal files, one can now build a format string and then use it in a subsequent write statement.

2.4.5   Exercises

1. Write a function that accepts a character string of arbitrary length as input and centers the text within the string. For example, if the input string is "aa   ", the result should be " aa ".

2. Write a program that reads a single line of text (up to, say, 128 characters long) and prints out the line with the words given in reverse order. For this purpose, words may be defined as strings of text delimited by any amount of blank space. For example, if the user types

    ```
    hello,    my name   is Bob
    ```

    then the response should be

    ```
    Bob is name my hello,
    ```

3. The intrinsic function **achar** returns the character corresponding to any position in the ASCII collating sequence. That is, if i is an integer between 0 and 127 inclusive, then **achar(i)** is a character string of length one containing the character in ASCII position i. Write a program that prints to the screen all the characters in the ASCII sequence (but note that some characters are nonprintable). From this output, identify the ASCII positions of the lowercase letters (a, ..., z), the uppercase letters (A, ..., Z), and the numerals (0, ..., 9).

4. The intrinsic function **iachar** accepts a character-string argument of length one and returns as an integer the ASCII position of that character. Using this function, write a procedure that converts all the lowercase letters within a string to uppercase or vice versa.

2.5   Additional Topics

2.5.1   Expressions with Mixed Types and Kinds

In general, one should be very cautious with integer division and with expressions or statements that mix integers with reals. Consider the simple test program below.

```
program test
   implicit none
   real :: x
   x = 2/3
   print *, x
end program test
```

Many novice programmers would be surprised by the result:

```
D:\jls\software\demos>test
  0.0000000E+00
```

In this example, Fortran interprets the literals 2 and 3 as integers of the default kind. It then computes 2/3 by integer division, which truncates the result to zero. Finally, the zero value is converted to a floating-point representation and stored in x. If 2/3 were changed to 2./3, 2/3., or 2./3., the result would become 0.666... because the literals 2. and 3. are interpreted as real numbers of the default kind (single precision).

Whenever an expression involving only integers is evaluated, Fortran returns the result as an integer. If an expression involves integers and reals of a given kind, the integers may be implicitly converted to reals of that kind, and the result is returned as real. If an expression involves reals of different kinds, those that are less precise may be converted to the higher precision, and the returned result is of the higher precision. If the expression is complicated, however, it may be partially evaluated before any conversion takes place. For example,

```
2/3 + 1.6
```

will evaluate to 1.6. Finally, when an assignment

variable = expression

is made, some conversion takes place if the value returned by *expression* does not match *variable* in type and kind. If *expression* is real and *variable* is an integer, the conversion will be made by truncation rather than rounding.

Large errors can arise in computational statements that mix data of different types or kinds because floating-point arithmetic is only approximate and because many integers have no exact representation in a floating-point model. For example, in

```
real :: a, b
```

```
integer :: i
a = 0.2
b = 3.8
i = a + b
```

the resulting value of i could be 3 or 4. For these reasons, mixed-type expressions should be avoided.

2.5.2   Explicit Type Conversion

To avert problems arising from mixing types and kinds, you can perform conversions yourself using the intrinsic functions **real** and **int**. If x is an integer or real variable, then **real(x,mykind)** returns a real value of the kind **mykind**. If the kind argument is omitted, the result will be a real value of the default kind. Similarly, **int(x,mykind)** converts the value in x to a **mykind** integer, and **int(x)** converts it to a default integer. Note that Fortran converts reals to integers by truncation toward zero, so **int(1.9989)** returns 1 and **int(-2.732)** returns -2. Both of these functions are elemental, so they can also be applied to arrays of arbitrary shape.

It is also important to understand how Fortran interprets literal constants in your source code. Numbers such as 0, 10, and -97 are regarded as integers. Anything involving a decimal point, such as 3.7, -9.0, and 6., is stored as a floating-point real number, but the kind may vary. Usually it will be of the default or single-precision kind, but if that model appears to be inadequate, the precision may or may not be increased. For example, some compilers will store 123456789.0 in double precision, noticing that this constant has more than six significant digits. However, others will disregard the extra digits and store it in single precision as approximately 1.23456×10^8. Putting the value into a double-precision variable, as in

```
double precision :: k
k = 123456789.0
```

or turning it into a double-precision named constant as in

```
double precision, parameter :: k = 123456789.0
```

may not recover the extra digits because the compiler may still interpret the literal with single precision before converting it to double precision. The best way to guarantee greater accuracy is to specify the literal in exponential notation using D rather than E. That is, the constant could be written as 1.23456789D8, 1.23456789D+8, or 1.23456789D+08. Exponential notation using an E, as in 1.4E-06, is generally interpreted with single precision, whereas D is interpreted as double precision.

If you still have any doubts about what your compiler is doing, the intrinsic function **kind** may be used to query the system to determine the actual **kind** parameter of any variable or constant. To see how your compiler handles the previous example, run this test program:

```
program test
   real :: a
   double precision :: b
   print *, "Single precision is kind", kind(a)
   print *, "Double precision is kind", kind(b)
   print *, "123456789.0 is kind", kind(123456789.0)
end program test
```

The way that literals are written may have important implications not only for accuracy but also for computational efficiency. For example, the statements

```
y = x**2
```

and

```
y = x**2.0
```

may seem equivalent, but a compiler implements them quite differently. The first simply computes **x*x**, whereas the second—raising a floating-point number to a floating-point power—is a far more complicated operation. Wherever possible, it is a good idea to use integer rather than floating-point exponents. It is also beneficial to express square roots as **sqrt(x)** rather than **x**0.5**, as the former allows the compiler to take advantage of highly efficient routines for computing square roots. More issues of numerical accuracy and efficiency will be taken up in Chapter 4.

2.5.3   Generic Procedures

Many Fortran intrinsic functions are generic. A generic function may accept arguments of different types and return values of different types. For example, the generic absolute value function **abs** may be applied to integer or real variables. If x is real, then **abs(x)** returns a real number of the same kind; if x is an integer, then **abs(x)** returns an integer of the same kind. Other commonly used generic functions include **sqrt, exp, log, log10, sign,**] **sin, cos,** and **tan.** All of these functions are elemental and can thus be applied to arrays.

When using these generic functions, it is helpful to check the language reference material supplied with the compiler to see what types of arguments are expected and what type of value will be returned; not doing so could produce unexpected results. For example, the generic square root function **sqrt** expects to operate on real numbers of various kinds but not on integers. Depending on the options of your compiler, **sqrt** may accept an integer argument and perform a type conversion to real or it may not.

That is, the expression `sqrt(4)` might possibly evaluate to 2.000... as intended, or it might produce an error when the program is compiled. To help ensure consistency of results across compilers, it would be wise to explicitly convert the argument with `real` in this case.

Fortran allows you to write your own generic procedures. This will be discussed in the next chapter, when we take up the subject of modules.

2.5.4 Don't Pause or Stop

We strongly recommend that you never use the Fortran statement `stop` within any subroutine or function, or even in a main program. This statement, along with its diabolical cousin `pause` (which was declared obsolescent in Fortran 90 and deleted from Fortran 95), appear in old-fashioned FORTRAN programs to halt execution in the event of an error. These statements may have unfortunate consequences for procedures embedded in DLLs or in COM servers. Rather than using `stop`, it's better to anticipate the possible errors that may arise and structure your procedures to exit gracefully if one occurs.

2.6 Additional Exercises

1. Explain why this snippet of Fortran code prints three different values:

   ```
   double precision :: x, a, b
   print *, 2./7.
   x = 2./7.
   print *, x
   a = 2.
   b = 7.
   x = a/b
   print *, x
   ```

   What result do you expect from this?

   ```
   real :: x
   x = 4.0
   print *, x**(1/2)
   ```

2. Write a procedure that takes as input an $n \times p$ data matrix X and a vector of weights $w = (w_1, \ldots, w_n)^T$ and computes the matrix of weighted sums of squares and cross products, $X^T W X$, where $W = \text{Diag}(w)$. Streamline the procedure by making use of the fact that $X^T W X$ is symmetric. Make the argument for w optional, so that if no weights are provided, the procedure computes $X^T X$ by default.

3. Write a procedure that accepts as input an $n \times p$ data matrix

$$
X = \begin{bmatrix}
x_{11} & x_{12} & \cdots & x_{1p} \\
x_{21} & x_{22} & \cdots & x_{2p} \\
\vdots & \vdots & \ddots & \vdots \\
x_{n1} & x_{n2} & \cdots & x_{np}
\end{bmatrix}
= \begin{bmatrix}
x_1^T \\
x_2^T \\
\vdots \\
x_n^T
\end{bmatrix}
$$

and computes the $p \times 1$ vector of sample means

$$
\bar{x} = \frac{1}{n} \sum_{i=1}^{n} x_i
$$

and the sample covariance matrix $S = k^{-1}A$, where

$$
A = \sum_{i=1}^{n} (x_i - \bar{x})(x_i - \bar{x})^T = X^T X - n\bar{x}\bar{x}^T
$$

and $k = n-1$ by default or $k = n$ by the caller's request. Use the one-pass method that updates $\bar{x}$ and A by incorporating rows of the data matrix one at a time. That is, letting $\bar{x}_{(i)}$ and $A_{(i)}$ denote the values of $\bar{x}$ and A based on the first i rows of the data matrix, initialize $\bar{x}_{(0)}$ and $A_{(0)}$ to zero and compute

$$
\bar{x}_{(i)} = \bar{x}_{(i-1)} + \frac{1}{i} \left(x_i - \bar{x}_{(i-1)} \right)
$$

$$
A_{(i)} = A_{(i-1)} + \frac{i-1}{i} \left(x_i - \bar{x}_{(i-1)} \right) \left(x_i - \bar{x}_{(i-1)} \right)^T
$$

for $i = 1, \ldots, n$.

4. Explain why the expression (2. == 2.D0) may evaluate to .true., whereas (exp(2.) == exp(2.D0)) may not. If you are not sure, write a test program to print out the value and kind of each subexpression.

5. Expanding e^x in a Taylor series about $x = 0$ and substituting $x = 1$ yields

$$
e = 1 + \frac{1}{1!} + \frac{1}{2!} + \frac{1}{3!} + \cdots.
$$

This suggests that the constant $e = 2.71828\ldots$ may be computed by this simple algorithm: set a, n, and e equal to 1, and repeat

$$
\begin{aligned}
a &= a/n, \\
e &= e + a, \\
n &= n + 1,
\end{aligned}
$$

until a is indistinguishable from zero. Write a Fortran program to compute e using the most precise floating-point model offered by your compiler.

6. If (X, Y) is uniformly distributed over the unit square—that is, if X and Y are independent $U(0, 1)$ random variates—then the probability of falling within the unit circle $(X - 1/2)^2 + (Y - 1/2)^2 \leq 1$ is $\pi/4$. Therefore, we can simulate the value of $\pi = 3.14159\ldots$ by generating a sample of n random points within the unit square, computing the proportion of the sample that falls within the circle, and multiplying by 4. Write a simple program that simulates the value of π in this manner for any given value of n.

7. Given a full-rank $n \times p$ data matrix X, there are many ways to transform it into another $n \times p$ matrix whose columns span the same linear space but are mutually orthogonal. One simple method is the Modified Gram-Schmidt (MGS) procedure (Golub and van Loan, 1996). This bit of Fortran code will decompose X into $X = QR$, where Q is an $n \times p$ orthonormal matrix and R is $p \times p$ upper-triangular. The X matrix is overwritten with Q.

```
double precision :: x(n,p), r(p,p)
integer :: n, p, j, k
r(:,:) = 0.D0
do j = 1, p
   r(j,j) = sqrt( sum( x(:,j)**2 ) )
   x(:,j) = x(:,j) / r(j,j)
   do k = (j+1), p
      r(j,k) = dot_product( x(:,j), x(:,k) )
      x(:,k) = x(:,k) - x(:,j) * r(j,k)
   end do
end do
```

Implement this MGS procedure in a function or subroutine, including a provision to prevent division by zero in case X is rank-deficient. Test your procedure on an X matrix filled with random numbers, and verify that $Q^T Q \approx I$ and $QR \approx X$.

8. Consider a two-way contingency table with elements x_{ij}, $i = 1, \ldots, r$, $j = 1, \ldots, c$. The estimated expected counts under a model of row-column independence are $\hat{x}_{ij} = x_{i+}x_{+j}/x_{++}$, where

$$x_{i+} = \sum_{j=1}^{c} x_{ij}, \quad x_{+j} = \sum_{i=1}^{r} x_{ij}, \quad x_{++} = \sum_{i=1}^{r}\sum_{j=1}^{c} x_{ij}.$$

A test for row-column independence is usually carried out by comparing the Pearson goodness-of-fit statistic

$$X^2 = \sum_{i=1}^{r}\sum_{j=1}^{c} \frac{(\hat{x}_{ij} - x_{ij})^2}{\hat{x}_{ij}}$$

or the deviance statistic

$$G^2 = 2 \sum_{i=1}^{r} \sum_{j=1}^{c} x_{ij} \log \frac{x_{ij}}{\hat{x}_{ij}}$$

to a chi-square distribution with $(r-1)(c-1)$ degrees of freedom. Write a procedure that accepts an assumed-shape two-way real array of nonnegative observed counts and computes the two-way array of estimated expected counts, X^2, G^2, and the degrees of freedom. When calculating G^2, handle observed counts of zero by treating $0 \log 0$ as zero. If any expected count is zero, then X^2 and G^2 are undefined and the procedure should signal a failure.

3

A Pseudo Object-Oriented Style

Subroutines and functions are excellent tools for breaking up complicated programs into smaller, more manageable units. After the introduction of C++ in 1983, however, a more sophisticated paradigm for modular development began to take hold, and the world of programming is now dominated by object-oriented languages and styles. A hallmark of object-oriented programming is encapsulation: grouping program variables and computational routines together intelligently and organizing them into logical units with only limited communication allowed between them.

Fortran 95 is not a true object-oriented language. The full object-oriented paradigm has been promised for Fortran 2003. Nevertheless, recent work by Akin (2003) and others has demonstrated that it is indeed possible to write structured Fortran 95 code that mimics many of the key qualities of object-orientedness. In this chapter, we first describe some important concepts from object-oriented programming. We then describe three new features of Fortran—modules, derived types, and pointers—and discuss their role in the development of self-contained software components. The chapter concludes with a detailed example of an object class, a generic error handler, which will be used to store and retrieve error and warning messages in all of our subsequent program examples.

3.1 Basic Concepts of Object-Oriented Programming

3.1.1 Objects and Classes

Loosely speaking, an object is a self-contained package of data bundled together with computational procedures that operate upon the data. More precisely, the archetype or design for the package is called a class, and an object is a particular instance or realization of the class created by a program. For example, consider the following variable declaration statement:

```
integer :: i, n
```

In this statement, `integer` can be regarded as the class, and `i` and `n` are two objects or instances of it.

Many objects of interest to a statistical programmer will be considerably more complicated than a single integer. For example, consider the classical weighted linear regression model

$$y = X\beta + \epsilon, \tag{3.1}$$

where y is a response vector of length n, X is an $n \times p$ matrix of covariates, β is a vector of p coefficients to be estimated, $\epsilon \sim N(0, \sigma^2 V)$, and

$$V^{-1} = W = \mathrm{Diag}(w_1, \ldots, w_n)$$

is a diagonal matrix of weights. An object designed to hold the input data for this analysis may contain two rank-one real arrays—one for holding the response variable and one for holding the weights—and a rank-two array for the covariates. The source code that defines the object class might be packaged together with routines that compute the weighted least-squares estimate

$$\hat{\beta} = (X^T W X)^{-1} X^T W y,$$

the estimated covariance matrix for $\hat{\beta}$, fitted values, residuals, and other diagnostics.

3.1.2 Properties

A property is an attribute of an object that may or may not be altered during program execution. In Fortran, even the simplest objects have a surprising number of properties. The properties of an array declared as

```
real(kind=our_dble), allocatable :: x(:)
```

include

- the rank (in this case, 1), type (integer), and kind (`our_dble`), which cannot be changed;

- the allocation status, size, shape, and lower and upper bounds, which can be set or changed by `allocate` and `deallocate`; and

- the floating-point data values held within the elements of the array, which can be set or changed by assignment statements.

From the programmer's perspective, each of these properties has important ramifications for how x may be used in a program. For example, in well-written code, any expression involving an element `x(i)` will never be evaluated unless x has been allocated and i lies between `lbound(x)` and `ubound(x)`. Depending on what purpose x serves within the program, only a few properties may be of interest to the user or influenced by what the user does. Recall, for example, the rank-one real array x used in the uniform generator program of Section 2.2.5 to hold random variates. The user, who may be entirely unaware of how Fortran works, is unlikely to care about the array's `kind` parameter or its lower and upper bounds; however, the program does require the user to set the array's size and then prints the data values stored in the array when the random generation is complete.

3.1.3   Put and Get

In the terminology of object-oriented programming, "put" is the process by which a property of an object is set or changed by an external entity, and "get" is the process by which an entity queries the object to determine the value of one of its properties. Consider the size of a rank-one real allocatable array x. We put the size by first deallocating x, if necessary, and then allocating it. We get the size by evaluating `size(x)`.

Whenever we design an object, we must first decide what the object's properties will be. Then, for each property, we must decide whether it will be a read-write property, whose value can be put or gotten by an external entity; a write-only property, whose value can be put but will never have to be gotten; or a read-only property, whose value can be gotten but not put. Read-only does not mean that the property's value will never change; rather, it means that a user of the object cannot change that property directly. For example, putting one read-write property could automatically change the values of many read-only properties. Restricting certain properties to be read-only is an excellent strategy for preventing the occurrence of unwanted events.

3.1.4   Methods and Constructors

A method is any operation or procedure applied to an object. Put is one special kind of method that sets a property, and get is another special kind of method that retrieves a property. Other methods may be more complicated, performing nontrivial computations and altering the values of one or more properties simultaneously. Consider again a rank-one real array x designed to hold random variates. It may be useful to make the size of x a read-write property and then create a method that fills x with random variates once its size has been set. The data values stored within x could then be made read-only, so that a user could never set them directly but only retrieve them after the random generation method has been called.

Another important type of method, called a constructor, defines the actions taken whenever a new instance of an object is created. The constructor determines the initial or default state of the object. For example, suppose that we design an object to hold the parameters $\theta = (\mu, \Sigma)$ of a multivariate normal distribution, where μ is a $p \times 1$ vector of means and Σ is a $p \times p$ positive-definite covariance matrix. One possible constructor for this object would set $p = 1$, $\mu = 0$, and $\Sigma = 1$ so that by default it specifies a univariate standard normal distribution. Another possible constructor would set $p = 0$ and leave the μ and Σ arrays in an unallocated state. In the latter case, one would probably want to create a put method for p that allocates μ and Σ and perhaps initializes μ to be a vector of zeroes and Σ to be an identity matrix.

3.1.5 Conceptualizing an Interface

The puts, gets, and other methods implemented for an object determine the manner in which the object may be used by the outside world. Taken together, these methods are called the interface. (This use of the term "interface," an important idea from object-oriented programming, should not be confused with the concept of an interface in Fortran, the specific rules by which dummy arguments in a function or subroutine are associated with their actual arguments. To remain consistent with the common terminology, we will use "interface" to refer to both, trusting that the reader will infer from the context which of the two is being discussed at any particular time.) By nature, statisticians who write programs or procedures may have a tendency to focus primarily on computational details and numerical methods, paying relatively little attention to the interface. In many cases, however, the quality of the interface is ultimately the most crucial factor in determining how useful the procedure will be.

Before embarking on a new project, it is an excellent idea to first conceptualize the interface. This exercise clarifies exactly what the software component will do and helps the programmer to remain focused throughout the development process.

Example: Testing Independence in a Two-Way Table

Imagine that you are given the task of writing a self-contained software component that performs the classical chi-square test for independence in an $r \times c$ contingency table. That is, given a table of nonnegative observed frequencies

$$\{x_{ij} : i = 1, \dots, r; \quad j = 1, \dots, c\},$$

your code will calculate the estimated expected frequencies under the model of row-column independence,

$$\hat{x}_{ij} = \sum_{j'=1}^{c} x_{ij'} \sum_{i'=1}^{r} x_{i'j} \left(\sum_{i'=1}^{r} \sum_{j'=1}^{c} x_{i'j'} \right)^{-1}, \quad (3.2)$$

the degrees of freedom

$$\nu = (r - 1)(c - 1), \quad (3.3)$$

the Pearson test statistic

$$X^2 = \sum_{i=1}^{r} \sum_{j=1}^{c} \frac{(\hat{x}_{ij} - x_{ij})^2}{\hat{x}_{ij}}, \quad (3.4)$$

and the asymptotically equivalent deviance statistic

$$G^2 = 2 \sum_{i=1}^{r} \sum_{j=1}^{c} x_{ij} \log \frac{x_{ij}}{\hat{x}_{ij}} \quad (3.5)$$

(see, e.g., Agresti, 2002). From an object-oriented standpoint, it makes sense to devise a new type of object—say, the `two_way_table` class—with the following properties and methods:

- `observed`, a write-only or read-write property that allows the user to input the matrix of x_{ij}'s;

- `run_chisquare_test`, a method that performs the computations in (3.2)–(3.5); and

- a set of read-only properties, `expected`, `df`, `Pearson`, and `deviance`, that allow the user to retrieve the matrix of $\hat{x}_{ij}$'s and the values of ν, X^2, and G^2.

An intelligent interface for the `two_way_table` class would perform necessary checks to anticipate and prevent illegal operations. For example:

- The put method for `observed` may test the input array to ensure that $r > 1$, $c > 1$, and $x_{ij} \geq 0$ for every i and j, returning an error if any of these conditions is violated.

- The `run_chisq_test` method may return an error if `observed` has not been put.

- The get methods for `expected`, `df`, `Pearson`, and `deviance` may return errors if `run_chisq_test` has not yet been invoked.

- Putting a new matrix of x_{ij}'s into `observed` may reset or nullify `expected`, `df`, `Pearson`, and `deviance`, so that `run_chisq_test` will have to be invoked again before any of these properties can be gotten.

3.1.6   Other Object-Oriented Concepts

Another key idea in object-oriented programming is polymorphism, whereby a single method or procedure can operate on objects of different classes with potentially different behavior in each case. A primitive kind of polymorphism has been implemented in Fortran. In Section 2.5.2, we discussed generic intrinsic functions, such as abs, that can be applied to integer and real variables of various kinds. More generally, Fortran programmers can write their own functions or subroutines that accept arguments of different types and perform type-specific operations on them. This feature, called overloading, can be implemented within a Fortran module using an interface block; this will be explained in the next section.

Inheritance is an asymmetric relationship between classes of objects in which one class shares all the properties and behavior of another class. Under this concept, "Class B inherits from Class A" means that B has all the properties and methods that apply to A and possibly more; A is then called the base or parent class, and B is called the derived or child class. For example, consider a class of objects wls_reg for performing weighted least-squares regression. If a class ols_reg for ordinary least-squares regression has already been developed, the programmer may create wls_reg by inheriting the properties of ols_reg and introducing an additional property, weights. Languages that support inheritance are convenient because they allow a programmer to create multiple classes of similar objects with minimal duplication of code.

Fortran 95 was never designed to support inheritance. It is possible, however, to write Fortran 95 code that implements both inheritance and runtime polymorphism, in which Fortran procedures dynamically distinguish objects of different types while a program is executing (Decyk, Norton, and Szymanski, 1998). In this book, we will not attempt to do either. We will, however, show the statistical programmer how to use modern Fortran to develop object classes with their own methods and interfaces, which may then become the building blocks of high-quality, reliable applications.

3.1.7   Exercises

1. Investigate the history of object-oriented programming. Which major programming languages implement all aspects of the object-oriented paradigm? Which languages are only partially object-oriented?

2. What is a method? Should every method alter one or more properties of an object? Explain.

3. What are the practical reasons for making a property of an object read-only? Would it ever be sensible to have a property that is write-only? Explain.

4. Choose a simple but nontrivial statistical procedure that is familiar to you. For example, if you have some background in survival analysis, the procedure could be computing the Kaplan-Meier estimate of a survivor function from a sample of failure times, some of which are right-censored. Conceptualize an object class for this procedure. Describe what actions will be taken by the constructor whenever an object of this class is created. List all the relevant properties of the object class, and decide for each property whether it will be read-write or read-only. List all the methods, and describe what each one will do, including any side effects.

3.2 Modules

3.2.1 What Is a Module?

A typical FORTRAN 77 application contained a main program that called various subroutines and functions. The source code for these subroutines and functions did not have to reside in the same file as the main program; it could be divided up into an arbitrary number of files that were compiled separately. After compilation, the multiple object-code files would be linked together to produce a single executable file, and the behavior of the resulting program would be no different than if all the source code had existed in one file.

With the advent of Fortran 90, a new organizational unit was introduced, called the module. A simple module may be nothing more than a collection of interrelated subroutines and functions, much like a source-code file in FORTRAN 77. But the module is capable of much more because it may also contain its own constants, variables, and derived-type definitions.

The essential syntax of a module is shown below.

```
module artichoke
   public
   ! declare constants, variables, and derived types here
   contains
   ! put subroutines and functions here
end module artichoke
```

The statement `contains` is optional and should only be included if subroutines and functions appear below it. The statement `public` declares that, unless otherwise specified, all of the contents of the module (e.g., named constants, variables, and procedures) will be accessible to any program unit that uses the module. Substituting `private` for `public` will declare that, unless otherwise specified, all the contents of the module will be hidden from all other program units whether they use the module or not. A module whose contents are all private can be of no use to other program units, so at least some features must be made public. On the other hand, automatically declaring all contents public can be undesirable because it makes the module less self-contained and increases the possibility that the module will interact with other program units in undesirable ways (e.g., conflicts arising because variables or subroutines in different modules have the same names). Artful design of modules must strike a careful balance between the desire to have unfettered communication with the outside world and the desire to minimize conflicts.

To use the public aspects of a module in another program unit, you need to include a **use** statement near the beginning of that unit. A module can be used by a program,

```
program cook_dinner
   use artichoke
   implicit none
```

by a subroutine or function,

```
subroutine make_antipasto(arg1, arg2, arg3)
   use artichoke
   implicit none
```

or by another module,

```
module cream_of_anything_soup
   use artichoke
   implicit none
```

In the last example, the public aspects of **artichoke** will be accessible to any function or subroutine contained within **cream_of_anything_soup**.

If a module is used by another program unit, that module must be compiled before compiling the other unit. For this reason, the source code for each module is typically placed in its own file, apart from the code for other modules or programs. Among other things, compilation allows Fortran to construct explicit interfaces for all of the module's procedures, which helps to ensure consistency between dummy and actual arguments.

3.2.2 *How Not to Use Modules*

One possible use for a module is to bundle together a group of variables that will be operated on by a function or subroutine. Consider the following modules designed to hold arrays used in weighted least-squares regression.

```
module wls_data
   public
   real(kind=our_dble), allocatable :: y(:), x(:,:), w(:)
end module wls_data
```

```
module wls_workspaces
   public
   real(kind=our_dble), allocatable :: xtwx(:,:), &
      xtwxinv(:,:), xtwy(:)
end module wls_workspaces

module wls_results
   public
   real(kind=our_dble), allocatable :: betahat(:), &
      cov_betahat(:,:), residuals(:), fitted_vals(:)
   real(kind=our_dble) :: sigma2hat
end module wls_results
```

Each of these modules creates variables that are global in the sense that they become accessible within any program or program unit that uses the modules. For example, consider this subroutine for calculating the symmetric matrix $X^T W X$:

```
subroutine make_xtwx
   use wls_data
   use wls_workspaces
   implicit none
   integer :: n, p, i, j, k
   n = size(x,1)
   p = size(x,2)
   allocate(xtwx, p, p)
   do j = 1, p
      do k = j, p
         xtwx(j,k) = sum( x(:,j) * w(:) * x(:,k) )
         xtwx(k,j) = xtwx(j,k)
      end do
   end do
end subroutine make_xtwx
```

In this fashion, one could write additional subroutines for computing the inverse of $X^T W X$, calculating $X^T W y$, and then multiplying them together to obtain the weighted least-squares estimate

$$\hat{\beta} = (X^T W X)^{-1} X^T W y.$$

A subroutine for fitting the regression might then look like this:

```
subroutine fit_wls
   call make_xtwx
   call make_xtwy
   call invert_xtwx
```

```
      call make_betahat
    end subroutine fit_wls
```

At first glance, this programming style seems attractive. The source code looks clean and compact because no arguments are being passed. However, this strategy violates a number of principles of good programming practice, and we do not recommend it. One reason is that, when data exist as global variables within a module, it becomes tedious to create or use multiple instances of them. For example, suppose we wanted to write a program that tests the fit of a null model against an alternative model that contains additional covariates; to keep the workspaces for the two models separate, we would have to create two copies of the `wls_workspaces` module in our source code and give them different names (e.g., `wls_workspace_null` and `wls_workspace_alt`). If we decided to make changes to the workspaces (e.g., adding additional arrays), we would need to make changes to the source code for both of the modules, and it would be impossible for one subroutine to use both workspaces because the arrays within them have been given the same names.

In our experience, we have found it undesirable for Fortran modules to contain any variables at all, whether public or private. Module parameters (i.e., named constants) are very useful, but module variables can be troublesome. In modern operating systems, it is possible to have multiple copies of the same program running simultaneously, and variables within modules can create unexpected conflicts that are difficult to diagnose and resolve. For these reasons, we will refrain from using variables in any of our modules from this point onward, and we urge our readers to do the same.

3.2.3   How to Use Modules

If we are not to use modules to hold data as variables, then how should we use them? One excellent way is to use modules to hold definitions for one or more public derived types and the subroutines and functions that operate on those derived types. Derived types—the subject of the next section—are custom-designed data structures of arbitrary complexity that can be instantiated within programs and passed as arguments to functions and subroutines. The definition of a derived type within a module determines its properties and forms the nucleus of an object class. Functions and subroutines that operate on the derived type become the gets, puts, and other methods that create an interface between the object and the outside world. Judicious use of `public` and `private` enables the objects to be created and manipulated by external programs while keeping the inner workings of the module hidden from them.

If you find this discussion to be somewhat abstract and confusing right now, don't worry. Gaining familiarity with object-oriented programming takes practice and time. Our approach of using Fortran modules to create object classes will be illustrated profusely throughout the rest of this book.

3.2.4 Generic Module Procedures

One benefit of modules is that they make it easy to create your own generic procedures. A generic procedure is actually a group of functions or subroutines that can be invoked by a common name. This feature, which is also called overloading, is illustrated by the following example.

Consider the subroutine below, which converts a single-precision real number to a left-justified character string.

```fortran
subroutine real_to_string( arg, str )
   implicit none
   real, intent(in) :: arg
   character(len=*), intent(out) :: str
   character(len=20) :: tmp  ! to prevent I/O errors
   write(tmp, *) arg
   str = adjustl(tmp)
end subroutine real_to_string
```

Here is another subroutine that does the same thing for an integer argument.

```fortran
subroutine int_to_string( arg, str )
   implicit none
   integer, intent(in) :: arg
   character(len=*), intent(out) :: str
   character(len=20) :: tmp  ! to prevent I/O errors
   write(tmp, *) arg
   str = adjustl(tmp)
end subroutine int_to_string
```

Because these two procedures are so similar, it may be helpful to refer to them by the same name, thereby reducing the number of procedures that we will have to remember in the future. To do this, we first place the two subroutines in a module below the **contains** statement. Then we write a simple **interface** block above the **contains** statement, like this:

```fortran
module string_conversion
   private ! by default
   public :: convert_to_string
   interface convert_to_string
      module procedure real_to_string
      module procedure int_to_string
   end interface
contains
```

The `real_to_string` and `int_to_string` subroutines are now private and cannot be called outside the module. However, both procedures are publicly available through the generic name `convert_to_string`. If we say

```
call convert_to_string( 3.14159, str)
```

or if we say

```
call convert_to_string( -897, str)
```

then Fortran automatically calls one procedure or the other, depending on the actual arguments that are supplied. By writing additional subroutines and listing them in the `interface` block, we can easily expand the capabilities of the generic procedure to handle double-precision reals, logical values, and so on.

An `interface` block can group either a set of subroutines or a set of functions, but we cannot have both subroutines and functions in the same block. When a generic procedure is called, Fortran chooses among the specific procedures by the nature of the nonoptional arguments. Therefore, the specific procedures must differ with respect to their arguments' number, type, kind, or rank.

3.2.5 Exercises

1. What are the most important differences between a module and an ordinary source-code file containing subroutines or functions?

2. A subroutine or function is considered to be external if its source code is located outside of the main program and is not a part of any other module or procedure. An external procedure does not have an explicit interface, so the compiler may not require the actual and dummy arguments to agree in type, kind, or rank.

   a. Show that it is possible (but not recommended) to write an external function whose dummy argument is a rank-one array but whose actual argument has rank two.

   b. Why do you think that we avoid the use of external procedures, placing them instead within modules? Give as many reasons as you can.

3. If module `mymod_a` is used by module `mymod_b`, then `mymod_b` cannot be used by `mymod_a`. Why not? (Hint: Think about the build process.)

4. Create a simple module that defines some useful double-precision real constants such as

$$\pi \approx 3.141592653589793,$$
$$e \approx 2.718281828459045.$$

Demonstrate how to use the constants from another procedure, module, or program.

5. Create a generic module procedure for computing variances. If the input is a rank-one real array of size $n \geq 2$, it should compute the sample variance of the n observations. If the input is a rank-two real array representing an $n \times p$ data matrix, it should compute the $p \times p$ covariance matrix. Use double-precision arithmetic throughout.

3.3   Derived Types

3.3.1   What Is a Derived Type?

In our discussion and programming examples thus far, all of our variables have either been scalars of one of the five atomic types (integer, real, complex, logical, or character) or arrays whose elements are one of those types. Arrays, although powerful, are limited in important respects. It is not possible to create a mixed-type matrix with one column containing reals and another containing integers. Nor is it possible to add properties to an array beyond the basic ones already provided by Fortran (kind, rank, shape, etc.). For example, there is no way to attach row or column names to a matrix. If there were no way to add properties to data structures, pseudo object-oriented programming in Fortran would be effectively impossible.

The derived type, first introduced in Fortran 90, is an arbitrary collection of scalar or array variables that have been glued together into a single entity. (Readers familiar with C will recognize that a derived type is basically the same as a "struct" in that language. Users of S-PLUS and R will recognize it as a list.) The concept is well-illustrated by a simple example. Suppose that you want to construct a database for holding vital information for people whom you know—names, addresses, telephone numbers, birthdates, and so on. Here is a definition for a derived type called `person_info_type` that can hold the desired data for a single person:

```
! change these parameters if you like
integer, parameter :: name_string_length = 30, &
   address_string_length = 40, phone_string_length = 20

type :: person_info_type
   character (len=name_string_length) :: last_name, &
      first_and_middle_names, nickname
   character (len=1) :: sex
   integer :: birth_day, birth_month, birth_year
   character (len=address_string_length) :: &
      email_address, snail_mail_address(4)
```

```
      character (len=phone_string_length) :: home, office, &
         mobile, fax
   end type person_info_type
```

Once this type definition has been included in a program, procedure, or module, we are free to create instances of the type and store data in them. The individual parts of the type, called components, are accessed by the component selector symbol %.

```
   type(person_info_type) :: person_info

   person_info%last_name = "Clemens"
   person_info%first_and_middle_names = "Samuel Langhorne"
   person_info%nickname = "Mark Twain"
   person_info%sex = "M"
   person_info%snail_mail_address(1) = "Woodlawn Cemetery"
   person_info%snail_mail_address(2) = "Elmira, NY"
```

 Style tip———————————————————————————

Notice that the new type was called **person_info_type**, whereas the specific instance of it was called **person_info**. We add the _type suffix to the name of every derived type to help maintain the crucial distinction between object classes and objects.

The components of a derived type may be scalars, arrays, pointers, and even derived types. For example, we may create a derived type to hold a birthdate,

```
   type :: birthdate_type
      integer :: day, month, year
   end type birthdate_type
```

use it within another type,

```
   type :: person_info_type
      character (len=info_string_length) :: last_name, &
         first_and middle_names, nickname
      character (len=1) :: sex
      type(birthdate_type) :: birth
         ! declare additional components here
   end type person_info_type
```

and access it like this:

```
   type(person_info_type) :: myself
   myself%birth%year = 1963
```

3.3.2   Using Derived Types

The variety of ways in which derived types may be used makes them extremely powerful. One may create arrays, even allocatable arrays, whose elements are instances of a derived type:

```
type(person_info_type), allocatable :: my_address_book(:)
allocate( my_address_book(100) )
```

Derived types and derived-type arrays may also be used as arguments to functions and subroutines. A function with a derived-type argument is shown below.

```
integer function find_person_by_nickname( nickname, &
      address_book, position) result(answer)
    ! Finds the position of the first person in address_book
    ! with a given nickname.
    character(len=*), intent(in) :: nickname
    type(person_info_type), intent(in) :: address_book(:)
    integer, intent(out) :: position
    ! locals
    integer :: i, n
    ! begin
    answer = 1
    position = 0
    n = size(address_book)
    do i = 1, n
       if( address_book(i)%nickname == nickname ) then
          position = i
          answer = 0
          exit
       end if
    end do
end function find_person_by_nickname
```

Notice that the dummy argument **address_book** is an assumed-shape array. Therefore, as discussed in Section 2.3.2, this function ought to be placed in a module. To compile successfully, we will have to place the definition of **person_info_type** in the same module, or in another module used by it, so that this type will be recognized. Here is the skeleton of a module that contains both the definition of the derived type and the function:

```
module mymod
    implicit none
    private ! by default
    public :: person_info_type, find_person_by_nickname
    integer, parameter :: name_string_length = 30, &
```

```
      address_string_length = 40, phone_string_length = 20
  type :: person_info_type
     character (len=name_string_length) :: last_name, &
        first_and_middle_names, nickname
     ! declare additional components here
  end type person_info_type
contains
  integer function find_person_by_nickname( nickname, &
     address_book, position) result(answer)
     ! put rest of statements here
  end function find_person_by_nickname
end module mymod
```

Any program or program unit that uses the module will be able to create instances of **person_info_type** and call **find_person_by_nickname**. The module's private parts—in this case, the string-length parameters—are hidden from the outside world, but the derived type and the procedure that uses it are accessible because both are public.

This strategy of placing the definition of a derived type, and any procedures that operate on the type, into a module is eminently sensible. In fact, it is the key idea of pseudo object-oriented programming in Fortran. We will use this technique extensively to develop self-contained code modules that can be shared among many applications yet are easy to maintain and extend without breaking the applications that use them.

To see how useful derived types can be in statistical programming, let us imagine a class of objects designed to hold the data for a weighted least-squares regression analysis, as we first described in Section 3.1.1. Suppose that we define a new type, **wls_data_type**, whose components include

- a rank-one real array y to hold the response variable,

- a rank-two real array x to hold the covariates, and

- a rank-one real array w to hold the weights.

In the future, we may wish to add additional features, such as character-string names for the response variable and the covariates; for now, however, let's consider only a bare-bones type that holds the numeric data. We will certainly want the new type to accommodate datasets of different sizes, allowing the number of observations and the number of covariates to vary. At this point, however, we have not yet discussed how to create a derived type whose components may vary in size. If we try to define the type in this way,

```
type :: wls_data_type
   integer :: n, p
   real :: y(n), x(n,p), w(n)
end type wls_data_type
```

the compilation will fail; the computer would have no way of knowing how much memory to set aside when creating an instance of the type. The definition

```
type :: wls_data_type
   real, allocatable :: y(:), x(:,:), w(:)
end type wls_data_type
```

will not work either because Fortran does not yet allow allocatable arrays to appear as components of derived types. (This feature has been promised for Fortran 2003, but it is not part of the 95 standard.) Every derived type created by Fortran 95 must have a physical size that can be discerned at the time of compilation, yet we clearly need components whose sizes may vary at run time. The solution to this problem is found in pointers; this will be discussed in Section 3.5.1.

3.3.3 Constructors and Default Initialization

A circle in two-dimensional space may be defined by its radius and by the x and y coordinates of its center. Here is a simple module that defines a derived type for a circle and a function that calculates its area:

```
module mymod
   type :: circle_type
      real :: center_x, center_y, radius
   end type circle_type
contains
   real function area(circle) result(answer)
      implicit none
      type(circle_type), intent(in) :: circle
      answer = 3.141593 * circle%radius**2
   end function area
end module mymod
```

If we declare an instance of a circle and immediately try to calculate its area,

```
type(circle_type) :: circle
print *, area(circle)
```

the result is unpredictable because the components of circle have not yet been defined. The problem is that we have not written a constructor. A constructor, as described in Section 3.1.4, determines the initial state of a newly instantiated object. For example, here is a procedure that initializes an instance to a unit circle centered at the origin:

```
subroutine initialize_circle_type(circle)
   implicit none
```

```
      type(circle_type), intent(out) :: circle
      circle%center_x = 0.
      circle%center_y = 0.
      circle%radius = 1.
   end subroutine initialize_circle_type
```

To use this constructor, we could place it within the module, make it public, and then invoke it from a program immediately after the instance is created.

```
   type(circle_type) :: circle
   call initialize_circle_type(circle)
```

Alternatively, we could assign initial values to all of the components within the definition of the derived type, like this:

```
   type :: circle_type
      real :: center_x=0., center_y=0., radius=1.
   end type circle_type
```

This latter method is called default initialization.

Unlike the initialization of local variables (Section 2.3.5), default initialization does not give components the **save** attribute. Default initialization affects the behavior of a derived type appearing as a dummy argument in a function or subroutines. If the dummy argument is **intent(out)**, the components will be given the initial values each time the procedure is called. For example, suppose that we create a derived type that defines a triangle in two-dimensional space. Then suppose we write a procedure that, given a triangle, finds the unique circle that circumscribes it. The procedure might begin like this.

```
   subroutine circumscribe(triangle, circle)
      implicit none
      type(triangle_type), intent(in) :: triangle
      type(circle_type), intent(out) :: circle
```

Whenever this procedure is called, the components of the dummy argument `circle` are initially set to the values given by the default initialization of `circle_type`. If default initialization were not used, then the components of `circle` would be undefined at the beginning of the procedure. In that case, we would need to be careful to explicitly set every component of `circle` to something within the procedure; otherwise, components of the corresponding actual argument would be undefined after the procedure is called.

 Style tip————————————————————————————————————

When creating a derived type, apply default initialization to each of the components.

3.3.4 Exercises

1. Create a derived type that can store a calendar date of the form "March 18, 2004" using a character string and two integers. Write a procedure that accepts this derived type as an argument and obtains the current date via a call to the intrinsic routine `date_and_time`. Write another procedure that writes the left-justified date to a character string.

2. The location of any point in two-dimensional space $\mathcal{R}^2$ can be represented by a pair of real numbers (x, y).

   a. Create a derived type that defines a point in $\mathcal{R}^2$-space. Then use this type to build additional derived types that define a circle (by its center and radius) and a triangle (by the locations of its three vertices).

   b. In two dimensions, a counterclockwise rotation of angle θ about a pivot point (x_0, y_0) transforms (x, y) to (x', y'), where

   $$
   \begin{aligned}
   x' &= x_0 + (x - x_0)\cos(\theta) - (y - y_0)\sin(\theta), \\
   y' &= y_0 + (x - x_0)\sin(\theta) + (y - y_0)\cos(\theta).
   \end{aligned}
   $$

   Write a procedure that rotates a point for any given angle and pivot. Overload the procedure to rotate circles and triangles as well.

3. Euclid (c. 365–275 B.C.) showed how to circumscribe a circle about any given triangle, as shown in Figure 3.1. Choose any two sides of the triangle and construct their perpendicular bisectors. The point at which these bisectors intersect is the center of the circle. (The bisector of the third side passes through this point as well.) The radius is simply the distance from this center to any of the three vertices. Implement Euclid's method as a Fortran procedure using derived types.

3.4 Pointers

3.4.1 Fear Not the Pointer

Please do not skip this section! Many novice programmers avoid pointers, finding them mysterious and frightening. This is unfortunate. Pointers, once understood, are an extremely valuable addition to the programmer's toolkit.

In other languages, pointers are flexible almost to the point of being dangerous; careless use of pointers leads to memory leakage and bugs that are

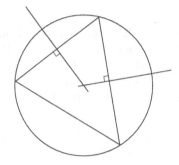

FIGURE 3.1. Euclid's method for circumscribing a circle about a triangle.

difficult to identify. In Fortran, however, use of pointers is much more limited. The architects of Fortran 90 and Fortran 95 enforced restrictions on pointers to enhance the ability of compilers to optimize code for faster execution. These restrictions also bring a degree of safety to the programmer because many pointer-related mistakes are detected during compilation.

3.4.2 Pointer Assignment

In other languages, a pointer can be regarded as a reference to a specific memory location. In Fortran, however, a pointer is best thought of as an alias, a portable name, that can be attached to any object of a given type. Consider the following snippet of code:

```
integer :: i
integer, target :: x(5,5)
integer, pointer :: a, b(:), c(:,:), d(:,:)
do i = 1, 5
   x(:,i) = i
end do
```

For the moment, let's ignore the word target in the second line; x is simply an array of integers whose present value is

$$\begin{bmatrix} 1 & 2 & 3 & 4 & 5 \\ 1 & 2 & 3 & 4 & 5 \\ 1 & 2 & 3 & 4 & 5 \\ 1 & 2 & 3 & 4 & 5 \\ 1 & 2 & 3 & 4 & 5 \end{bmatrix}.$$

The declaration of c as an integer pointer of rank two does not create an array called c. Rather, c is a name that can be assigned to an already existing rank-two integer array. Pointer assignment is accomplished by the symbols =>, like this:

```
c => x
```

After this statement is executed, c is another name for x. The statements

```
print *, c(2,3)
print *, size(c)
```

will display the results 3 and 25, and

```
c(2,:) = 7
```

will change the value of x to

$$\begin{bmatrix} 1 & 2 & 3 & 4 & 5 \\ 7 & 7 & 7 & 7 & 7 \\ 1 & 2 & 3 & 4 & 5 \\ 1 & 2 & 3 & 4 & 5 \\ 1 & 2 & 3 & 4 & 5 \end{bmatrix}.$$

A pointer is different from an allocatable array. If we had declared

```
integer, allocatable :: c(:,:)
```

then the statements

```
allocate( c( size(x,1), size(x,2) ) )
c = x
```

would have made c an independent copy of x occupying its own section of memory. As a pointer, however, statements involving c access the same memory locations as statements involving x; c and x refer to a single array.

Before a variable can be "pointed to," it must be given the target attribute; this is why

```
integer, target :: x(5,5)
```

had to be used. A pointer may point to any variable with the target attribute, provided that the target is of the same type, kind, and rank. For example, the statement

```
b => x
```

is not allowed because x has rank two but b was declared to have rank one. However, the name b may be assigned to a section of x, provided that the section has rank one:

```
b => x(1,:)
```

Similarly,

```
a => x(4,5)
```

is also allowed. A pointer may also be pointed to another pointer, in which case it is actually directed to the latter's target; for example,

```
c => x
d => c
```

and

```
c => x
d => x
```

do precisely the same thing.

3.4.3 Pointer Status

Once a pointer has been directed to a target, it is said to be associated. A pointer can be made to point to nothing, in which case it is said to be disassociated or null; this is accomplished by the **nullify** command:

```
nullify(a)
```

If a pointer has been nullified, we should not attempt to access its contents until it is assigned to another target; doing so, as in

```
nullify(a)
print *, a
```

may cause a crash. To prevent such mistakes, we may test the status of a pointer using the intrinsic inquiry function **associated**. The expression **associated(a)** evaluates to **.true.** if **a** is associated with a target and **.false.** if **a** is null.

In addition to being associated or nullified, a pointer's status may be undefined. A pointer declared at the beginning of a program, as in

```
program test
   real :: x
   integer, pointer :: a
```

remains undefined until it is either assigned to a target or explicitly nullified. (This behavior is no different from any other kind of variable; in the example above, x is also undefined until it is given a value.) Leaving a pointer dangling in an undefined state is extremely undesirable because any operation upon it—even testing its status with the **associated** function—could cause a crash. For this reason, programmers are strongly encouraged to nullify all pointers at the beginning of a program. In Fortran 90, this had to be done through **nullify** statements. Repeated use of **nullify** at the beginning of a program or procedure is simple enough, but it can also be somewhat tedious. To simplify matters, a feature was added in Fortran 95 to declare and nullify new pointer variables at the same time. The new feature is the **null()** constructor, which can be used like this:

```
      real, pointer :: x => null()
      integer, pointer :: c(:,:) => null(), b(:) => null()
```

3.4.4 Pointer Allocation

The `allocate` statement may be applied to a pointer, but the effect is somewhat different from when it is applied to an allocatable array. Suppose we declare a pointer to a real value,

```
real, pointer :: x
```

and then allocate it within a program:

```
allocate(x)
```

Three actions will be taken. First, if x happens to be associated with a target, it is disassociated from that target. Second, the processor sets aside a section of the heap (currently unused memory) large enough to hold a single real value. Third, x is pointed to that new section of memory, so that x can now be used to store a datum.

In a similar fashion, we can also allocate array pointers. If x is declared as

```
integer, pointer :: x(:,:)
```

then the statement

```
allocate( x(100,2) )
```

disassociates x from any current target, sets aside a section of the heap large enough to hold a 100×2 array of integers, and points x to those new memory locations.

Suppose that we apply `allocate` to a pointer that is presently assigned to an allocatable array:

```
integer, allocatable, target :: y(:,:)
integer, pointer :: c(:,:) => null()
c => y
allocate( c(2,2) )
```

In this case, y is not allocated; rather, c is first disassociated from y, and then c is allocated, so that y and c become two different arrays. This is a rather bad practice and should be discouraged for reasons that we now describe.

3.4.5 Pointer Deallocation

If c is a pointer that was previously allocated, then

```
deallocate(c)
```

returns the target section of memory to the heap and nullifies c. When deallocating pointers, we need to be careful because `deallocate` should only be applied to a pointer

- that is currently associated with a target, and

- whose target was previously created by an `allocate` statement.

We should never, for example, attempt to deallocate a pointer that is null or undefined. Nor should we attempt to deallocate a pointer that is currently pointing to an allocatable array or an array section.

Violations of the first kind are easily preventable if we (a) nullify every pointer with `null()` at the outset and (b) always test the association status with `associated` before deallocating. But the second kind of violation— trying to deallocate a pointer whose target goes by another name—is more insidious. To prevent those mistakes, it is an excellent practice to draw a sharp distinction in any program between two kinds of pointers: those that will be pointed only to targets that already go by other names, and those whose targets will only be created by allocation. If we decide beforehand never to apply `allocate` or `deallocate` to a pointer designated for the first purpose, and never to point the second kind of pointer to a preexisting variable, array, or array section, then we will protect ourselves from trouble.

3.4.6 Memory Leaks

What happens if we allocate repeatedly? Consider this:

```
allocate( x(100, 2) )
allocate( x( 40, 7) )
allocate( x(  5,23) )
```

If x were an allocatable array, then the program could crash at the second line; an allocatable array must be deallocated before it can be allocated again. If x is an array pointer, however, the program does not crash; the second line disassociates x from the 100×2 array and points it to a new section of size 40×7, and the third line disassociates x from the 40×7 array and points it to a new section of size 5×23. After the third allocation, the 100×2 and 40×7 sections are inaccessible. Data can no longer be stored in or retrieved from them, nor can they be deallocated and returned to the heap, because they no longer have names.

This undesirable effect is called memory leakage. Leaky programs can be inefficient, occupying more space than necessary. If pointers are repeatedly allocated within an iterative procedure, the program may eventually run out of memory and crash. Leaks can be stopped by judiciously checking the status of pointers, putting a statement such as

```
if( associated(x) ) deallocate(x)
```

before each allocation.

3.4.7 Exercises

1. Explain what action is taken by each line of code below, and identify the state of x, y, p, and q at each step.

   ```
   integer, target :: x = 5, y = 0
   integer, pointer :: p => null(), q => null()
   p => x
   y = p
   q => y
   q = 7
   p => q
   allocate(q)
   q => p
   ```

2. Suppose that a pointer appears as a local variable within a procedure. If the pointer is allocated in the procedure but never explicitly deallocated or nullified, what is its status the next time the procedure is called? Does this depend on whether the null() constructor was used? (If you are not sure, write a test program.)

3. Fortran allows you to write a procedure in which a dummy argument is not a pointer but the corresponding actual argument is. To see how this can be useful, apply a simple matrix operation (e.g., transpose) to a rank-two layer of a rank-three array.

4. Fortran does not allow you to create an array whose elements are pointers. It is possible, however, to define a derived type whose only component is a pointer and then create an array of that type. Demonstrate with an example.

3.5 Why We Need Pointers

3.5.1 Pointers in Derived Types

At this point, the reader should understand what a pointer is and how to avoid common pointer-related mistakes and bad practices. However, we have not yet presented any compelling reasons why pointers are necessary or even helpful.

Is it true that anything one can do with pointers can also be done just as effectively—and perhaps more safely—with ordinary variables, arrays, and allocatable arrays? Not at all. One very important reason for using pointers is that they allow us to create derived types whose components vary in size. In Section 3.3, we were wondering how to design an object to hold the data needed for weighted least-squares regression. The only way to do this in Fortran 95 is to use pointers within a derived type:

```fortran
type :: wls_data_type
   real, pointer :: y(:) => null(), x(:,:) => null()
   logical :: weights_present = .false.
   real, pointer :: w(:) => null()
end type wls_data_type
```

We create an instance of this type within a program simply by declaring it:

```fortran
type(wls_data_type) :: data
```

 Style tip ───

Using the null() constructor within the type definition ensures that the pointers within any new instance of the type are nullified.

───

Once we have determined the number of cases n and the number of covariates p for the current dataset, we may allocate the components of· data, like this:

```fortran
if( associated( data%x) ) deallocate( data%x )
allocate( data%x(n,p) )
```

3.5.2 Pointers as Dummy Arguments

A second reason why we need pointers is to pass them as arguments to functions or subroutines that may alter their size or shape.

Continuing with the weighted least-squares example, suppose that we define a derived type called `wls_results_type` to hold the results of the regression analysis. The components of this type may include the estimated coefficients $\hat{\beta}$, the estimated covariance matrix for $\hat{\beta}$, fitted values, and residuals:

```fortran
type :: wls_results_type
   real, pointer :: beta(:) => null(), &
         cov_beta(:,:) => null(), fitted(:) => null(), &
         residuals(:) => null()
end type wls_results_type
```

It would not make sense to allow a user to assign values to the components of the `wls_results_type` directly because these values are computed from the data by the regression procedure. Rather, the user should be allowed only to retrieve values of these components after the regression procedure has been run. In the language of object-oriented programming, the properties of `wls_results_type` should be read-only. To make them read-only, two things must be done. First, `wls_results_type` should be defined in a module, and its contents should be kept private so that its components may not be accessed directly outside the module. This can be done by placing a `private` statement in the derived type definition, like this:

```
type :: wls_results_type
   private
   real, pointer :: beta(:) => null(), &
```

Second, we should place in the same module a set of public procedures for getting the values of the components. A function for getting the estimated coefficients $\hat{\beta}$ is shown below.

```
integer function get_wls_beta(beta, results) result(answer)
   ! For getting the estimated coefficients from WLS.
   ! Returns 0 if successful, 1 if error.
   implicit none
   ! arguments
   real, pointer :: beta(:)
   type(wls_results_type), intent(in) :: results
   ! locals
   integer :: p
   ! begin
   answer = 1
   if( .not.associated( results%beta ) ) goto 500
   p = size(results%beta)
   if( associated(beta) ) deallocate(beta)
   allocate( beta(p) )
   beta(:) = results%beta(:)
   ! normal exit
   answer = 0
   return
   ! error trap
500   continue
   ! Later we ought to add something here to report a
   ! meaningful error message, telling the user
   ! that there are no estimated coefficients to
   ! be gotten.
end function get_wls_beta
```

Notice that the dummy argument `beta` is a pointer to a rank-one real array. This means that the user does not need to worry about the length of the coefficient vector in advance; he or she simply passes a rank-one real pointer to the procedure, and the pointer is automatically allocated to the correct size.

In this example, it seems logical to give the dummy argument **beta** the attribute **intent(out)**. However, Fortran does not allow us to declare intent for pointer dummy arguments; doing so would create confusion because it would be unclear whether this attribute refers to the pointer or to its target.

3.5.3   Recursive Data Structures

A third reason why we need pointers is to create recursive data structures, such as linked lists and binary trees. Recursive data structures tend to be an unpopular topic among students in introductory programming courses. Nevertheless, these structures are quite useful. In statistical applications, linked lists and trees provide efficient ways to tabulate and sort data, to find medians and other quantiles, and to accumulate data incrementally without having to repeatedly resize arrays and copy their contents.

To implement a recursive data structure in Fortran, we need to create a derived type that "contains itself" in the following sense: one or more components of the type are pointers to objects of that type. To see why this can be useful, suppose that we want to create a data structure to store text message lines that can grow during program execution to hold as many additional strings of text as necessary. One way to accomplish this is to create an allocatable array to hold the first message line:

```fortran
integer, parameter :: msg_width = 70
character (len=msg_width), allocatable :: msgs(:)
allocate( msgs(1) )
```

To add additional lines, we would need to copy the messages to a temporary array, resize the original one, and transfer the contents back. All of this allocation, deallocation, and copying makes the process quite inefficient. A much better strategy is to implement the message handler as a linked list.

To build a linked list, we first create a derived type to serve as a single node. This type has two components: a text string and a pointer to another instance of the type.

```fortran
type :: msg_line_type
    ! a single node in the linked list
    character (len=msg_width) :: line = ""
    type(msg_line_type), pointer :: next => null()
end type msg_line_type
```

By declaring one instance, we create the first node; additional nodes are introduced by pointer allocation. For example,

```fortran
type(msg_line_type) :: first_line
allocate( first_line%next )
```

creates two linked nodes, as shown in Figure 3.2. To build a longer linked

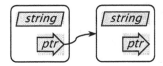

FIGURE 3.2. Two linked instances of a recursive derived type.

list, we can declare another pointer of the same type

```
type(msg_line_type), pointer :: current_line
```

and allocate it repeatedly, always keeping it pointed to the last node in the list:

```
current_line => first_line
do i = 1, n
   allocate( current_line%next )
   current_line => current_line%next
end do
```

3.5.4 Procedures for Linked Lists

To make the linked list easier to use, it helps to create a derived type that "contains" the entire list. An object of this new type can then be passed as an argument to procedures that will insert additional nodes, print the contents of the list, and so on. Let's define a `msg_list_type`, whose elements include pointers to the first and last nodes, along with a logical variable to indicate whether the list is empty:

```
type :: msg_list_type
   ! structure to contain the entire list
   logical :: msg_present = .false.
   type(msg_line_type), pointer :: head => null(), &
       tail => null()
end type msg_list_type
```

A new instance of this type will be an empty list. Here is a procedure for adding a new message line to the list:

```
subroutine insert_msg_line( text_line, msg_list )
   ! inserts a new message line at the end of the list
   implicit none
   character(len=*), intent(in) :: text_line
   type(msg_list_type), intent(inout) :: msg_list
   if( .not. msg_list%msg_present ) then
      ! begin a linked list
      allocate( msg_list%head )
```

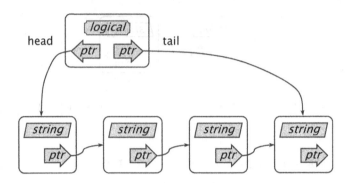

FIGURE 3.3. A linked list for storing text message lines.

```
        msg_list%tail => msg_list%head
        msg_list%head%line = text_line
        msg_list%msg_present = .true.
     else
        ! add a node to the list
        allocate( msg_list%tail%next )
        msg_list%tail => msg_list%tail%next
        msg_list%tail%line = text_line
     end if
  end subroutine insert_msg_line
```

If we create a new instance of the `msg_list_type` and call the procedure four times,

```
  type(msg_list_type) :: msg_list
  call insert_msg_line( "Good morning.", msg_list )
  call insert_msg_line( "How are you today?", msg_list )
  call insert_msg_line( "I'm so glad to see you.", msg_list )
  call insert_msg_line( "That's all for now.", msg_list )
```

then the list will have four nodes, as shown in Figure 3.3.

The next procedure prints all of the message lines to the screen in the order in which they were stored.

```
  subroutine print_msg_contents( msg_list )
     ! prints the text message lines to standard output
     implicit none
     type(msg_list_type), intent(in) :: msg_list
     type(msg_line_type), pointer :: current_line
     if( msg_list%msg_present ) then
        current_line => msg_list%head
        do
```

```
            print "(A)", current_line%line
            if( .not.associated(current_line%next) ) exit
            current_line => current_line%next
        end do
     end if
  end subroutine print_msg_contents
```

Finally, we need a procedure to delete all messages from the list. Merely nullifying the head and tail pointers is unacceptable because that will cause a memory leak; we need to explicitly deallocate each node.

```
subroutine delete_msg_contents( msg_list )
   ! deletes all messages from the list
   implicit none
   type(msg_list_type), intent(inout) :: msg_list
   type(msg_line_type), pointer :: current_line
   if( .not. msg_list%msg_present ) return
   do
      current_line => msg_list%head
      msg_list%head => msg_list%head%next
      deallocate( current_line )
      if( .not.associated(msg_list%head) ) exit
   end do
   nullify(msg_list%tail)
   msg_list%msg_present = .false.
end subroutine delete_msg_contents
```

Writing procedures for linked lists takes a little practice and requires a good understanding of pointers and derived types. Notice, however, that once these procedures are written, they are extremely easy to use:

```
type(msg_list_type) :: msg_list
call insert_msg_line( "Hello, world.", msg_list )
call print_msg_contents( msg_list )
call delete_msg_contents( msg_list )
```

From the user's perspective, the derived type `msg_list_type` and its procedures have become a "black box." The user may remain blissfully ignorant of how they work and does not even need to know that the `msg_list_type` is a linked list. This illustrates perhaps the greatest advantage of the object-oriented approach to programming. Details about the inner workings of the objects are kept hidden from the outside world and may be enhanced or changed in the future without requiring changes to programs and procedures that use them. To make sure that the details are kept hidden, we need to encapsulate the type definition and its procedures in a Fortran module, making some of the module's contents public but keeping others private.

3.5.5 Exercises

1. Using pointers, write a subroutine called `matrix_transpose` that accepts a rank-two real array (not necessarily square) as its only argument and transposes it. Can you do this without pointers? Explain.

2. In Section 2.3.3, we presented a function that calculates $y = \sqrt{x}$ to single precision by a Newton-Raphson algorithm. Using pointers, modify this procedure so that it optionally returns the number of iterations performed and the estimated value of y at each iteration.

3. An object class for performing the standard chi-square test for independence in an $r \times c$ contingency table was described conceptually in Section 3.1.5. We can now begin to implement this object class in Fortran using modules, derived types, and pointers.

    a. Create a derived type called `two_way_table_type` whose components include `observed`, `expected`, `df`, `Pearson`, and `deviance` (note that some of these must be pointers). Provide default initialization for all components. Place the derived type within a module. Declare the derived type to be `public` but keep all of its components `private`.

    b. Write a public procedure `put_observed` for entering a table of observed counts x_{ij} into the derived type. The dummy arguments for this procedure should include a `two_way_table_type` and a pointer to a rank-two real array containing the x_{ij}'s.

    c. Write a public procedure `run_chisquare_test` that performs the computations shown in Equations (3.2)–(3.5) on an instance of the `two_way_table_type`.

    d. Write public procedures for getting the values of the read-only properties `expected`, `df`, `Pearson`, and `deviance`.

    e. Write a simple test program that calls each of these module procedures.

4. Recursive data structures naturally lend themselves to recursive procedures. Create a derived type that can serve as a node in a linked list. Write recursive subroutines for adding another node to the end of the list and for deleting the entire list.

5. Implement a personal telephone directory as a linked list. Each node in the list should include a left-justified character string for the person's name, the telephone number, and a pointer to the next node. Write procedures for adding a new node, for printing the entire list, and for deleting the list.

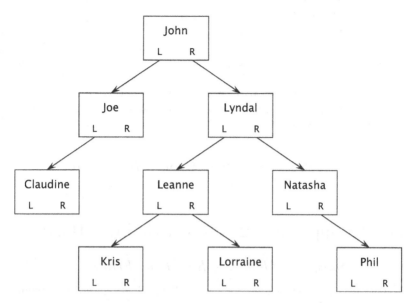

FIGURE 3.4. Binary tree containing an alphabetically ordered list of names.

6. Modify the linked list from the previous exercise so that entries are stored in alphabetical order with respect to name. That is, instead of inserting new nodes at the end of the list, insert them at an appropriate place to maintain the ordering. For simplicity, determine alphabetical order by comparing character strings with the intrinsic functions llt, lle, lgt, or lge.

7. A binary tree is a recursive data structure in which each node contains pointers to two additional nodes. The left-hand pointer is directed toward a node that is (in some sense) less than the current node, and the right-hand pointer is directed at a node that is greater. A simple binary tree containing an alphabetically ordered list of names is shown in Figure 3.4.

   a. Create a derived type called **tree_node_type** that represents a single node in the binary tree shown in Figure 3.4.

   b. Write a recursive subroutine that adds a new name to a tree by allocating a node at the appropriate place. This subroutine should have two dummy arguments: a node and a character string.

   c. Write a recursive procedure that prints all of the tree's names in alphabetical order.

    d. Write a recursive procedure for destroying (i.e., deallocating) the entire tree.

    e. Write a test program that adds the names shown in Figure 3.4 to the tree in an arbitrary order, prints out the ordered list, and then destroys the tree.

8. Using a binary tree, write a procedure that tabulates numeric data. That is, given a rank-one array of real numbers $x_1, x_2, \ldots, x_N$, find the unique values

$$x_1^* < x_2^* < \cdots < x_n^*$$

($n \leq N$) and their associated frequencies $f_1, f_2, \ldots, f_n$.

3.6 Example Module: A Generic Error Handler

3.6.1 Strategy for Managing Run-Time Errors

Except for the most trivial examples, all applications crash at countless points during their development, testing, and use. Just a few reasons are:

- attempting to open a file for read-access when the file does not exist;

- attempting to read past the end of a record (EOR) or the end of a file (EOF);

- attempting to allocate an array that is already allocated; and

- attempting to invert a matrix that is apparently singular.

Sometimes the problem arises because someone attempts to use the program incorrectly or in a manner that the author does not anticipate. At other times, the crash reflects a genuine bug that needs to be fixed within the source code.

All too often, programmers think, "I'll just get the program working for now, and put in some error-handling features later." Of course, "later" never arrives because run-time errors cannot be handled by adding a few lines of code here and there. Effective error management requires careful planning and influences the design of procedures at every level. Recognizing that run-time errors are inevitable, one should develop a coherent strategy for handling them and apply it consistently from the start.

Ideally, we would like to isolate any problem before a run-time error (such as an arithmetic exception) occurs, diverting the flow of control to prevent a crash. The subject of anticipating exceptions is discussed in the next chapter. (See section 4.1.4.) We also want to report meaningful messages to users at run time so that they may understand what happened and take corrective measures. The error messages should also be detailed enough to facilitate debugging, helping the developer or support staff to quickly locate the portion of source code where the problem occurred. Finally, we want the error-handling facility to be simple to use, to avoid tedious repetition, and to keep the code clean and compact.

As mentioned in Section 2.5.4, the statement **pause** is obsolete and no longer acceptable, and the use of **stop** is strongly discouraged. These statements halt the flow of control as soon as they are encountered, preventing the program from choosing alternative actions more appropriate to the context. Consider a procedure for matrix inversion. If the procedure encounters a singular matrix, halting the program may be reasonable in some situations, but in others we may want to continue execution, switching to a method capable of computing a generalized (e.g., Moore-Penrose) inverse. In general, when an error occurs within a procedure, the ultimate decision about what action to take should not be made within the procedure itself but should be deferred to the program unit that calls it.

We now develop a unified strategy for handling errors that will be used in programs throughout the rest of the book. Our strategy has two crucial features. First, whenever an error occurs within a procedure, control is returned to the program unit that calls it. Second, we create a class of objects for storing a sequence of text error messages. Whenever an error occurs, we store an informative message that includes a precise description of the error to help the program user to understand what happened. In addition, we also store the name of the program unit in which the error occurred, along with the names of any higher-level units that called it, so that a traceback of the calling sequence is available to the programmer. The code that defines the object class for holding error messages is placed in a module, along with its methods, illustrating the major principles of pseudo object-oriented design.

3.6.2 Structure of the Module

Our module, which we call **error_handler**, will allow us to store an arbitrary number of error message text lines and retrieve them at any time. As in the previous section, we will implement the messaging system as a linked list. Details of how the list works will be hidden from the outside world; from the user's perspective, the entire error messaging system will act like a "black box."

The heart of the module is **error_type**, an object class for holding the linked list of messages. The type itself is public so that it can be used by the outside world, but its contents are kept private. All communication between an instance of **error_type** and the outside world is handled by public functions and subroutines. The module without its procedures is shown below.

```
───────────────── error_handler.f90 ─────────────
!######################################################################
module error_handler
   ! Generic error message handler for both console and non-console
   ! applications. The routines in this module do not halt program
   ! execution; they merely store error messages for subsequent
   ! retrieval.
   ! Written by J.L. Schafer, 4/19/02; revised 6/03/03
   implicit none
   private ! by default
   ! declare public types
   public :: error_type
   ! declare public subroutines and functions
   public :: err_reset, err_handle, err_msg_present, err_get_msgs
   ! Parameters private to this module
   integer, parameter :: &
        ! max width of any single error message line
        err_msg_width = 70
   !######################################################################
   type :: msg_line_type
      ! Private type for a single node in the linked list
      sequence
      character (len=err_msg_width) :: line = ""
      type(msg_line_type), pointer :: next => null()
   end type msg_line_type
   !######################################################################
   type :: error_type
      ! Public type for holding a linked list of messages
      sequence
      private ! contents of this type are private to this module
      logical :: msg_present = .false.
      type(msg_line_type), pointer :: head => null(), tail => null()
   end type error_type
      !######################################################################
contains
      !######################################################################
      ! all subroutines and functions will be placed here
      !######################################################################
end module error_handler
   !######################################################################
```

 Style tip ───

The **private** statement near the beginning of the module makes everything defined in the module—parameters, derived types, functions, and subroutines—private unless otherwise stated. For pseudo object-oriented programming, it's usually a good idea to use the **private** statement and explicitly list the public parts below it.

In this module, notice that `msg_line_type` is private; this derived type is used to set up the data structure within the module but will not be known to the outside world. Similarly, the contents of `error_type` are kept private even though the type itself is public.

Notice also the **sequence** statement in the definition of each derived type. Using **sequence** causes the data within a type to be stored in memory in the same sequence as the components are declared. This statement does not affect how the type will be used in other parts of the program, but it does facilitate the transfer of data between Fortran and applications written in other languages. Using **sequence** will be necessary when creating COM servers as described in the later chapters of this book. We will use this directive in all of our derived-type definitions from now on.

3.6.3 Module Procedures

All access to the error-handling system is managed by four public procedures:

- `err_handle`, which appends an error message to an `error_type` object;

- `err_msg_present`, which queries the object to see if any messages are currently stored;

- `err_get_msgs`, which extracts all of the messages stored in the object and places them in a character-string buffer, allowing them to be easily printed to the screen or written to a text file; and

- `err_reset`, which deletes all messages currently stored in the object, returning it to its initial null state.

These public procedures, and any private procedures used by them, appear in the module below the **contains** statement. First, let's look at `err_reset`, which is essentially identical to the procedure in Section 3.5.4 for emptying a linked list.

```
───────────────── error_handler.f90 ─────────────────
!##################################################################
subroutine err_reset( err )
   ! Public: deletes all messages from the list
   implicit none
   type(error_type), intent(inout) :: err
   type(msg_line_type), pointer :: current_line
   if( .not. err%msg_present ) return
   do
       current_line => err%head
       err%head => err%head%next
```

```
      deallocate( current_line )
      if( .not.associated(err%head) ) exit
   end do
   nullify(err%tail)
   err%msg_present = .false.
end subroutine err_reset
!################################################################
```

The next function returns .true. if a message is present and .false. otherwise.

```
──────────────── error_handler.f90 ────────────────
!################################################################
logical function err_msg_present(err)
   ! Public: Queries the error_type to see if a message is present
   implicit none
   type(error_type), intent(inout) :: err
   err_msg_present = err%msg_present
end function err_msg_present
!################################################################
```

Next, we have a routine for appending a single line to the list. This procedure is private and will be used by `err_handle`.

```
──────────────── error_handler.f90 ────────────────
!################################################################
subroutine insert_msg_line( text_line, err )
   ! inserts a new message line at the end of the list
   implicit none
   ! declare arguments
   character(len=*), intent(in) :: text_line
   type(error_type), intent(inout) :: err
   ! begin
   if( .not. err%msg_present ) then
      ! begin a linked list
      allocate( err%head )
      err%tail => err%head
      err%head%line = text_line
      err%msg_present = .true.
   else
      ! add a node to the list; point tail to the new node
      allocate( err%tail%next )
      err%tail => err%tail%next
      err%tail%line = text_line
   end if
end subroutine insert_msg_line
!################################################################
```

Now let's consider `err_handle` itself. This subroutine has two required arguments: an `error_type` object and an integer code for reporting a variety of common errors. It also has a number of optional arguments through which the programmer can report very specific and useful information depending on the type of error. With input/output errors, for example, one may report the name of the file in question and perhaps even the line number in the file where the problem arose. One may also supply the name of the subroutine or function where the error occurred. If we adhere to the practice of always returning control to the calling procedure if an error occurs and always reporting the name of the program unit to the error handler, then a complete traceback of the calling sequence will be stored for debugging purposes. This subroutine, whose major parts are shown below, also provides a good example of the **case** construct described in Section 2.1.8.

```
───────────────── error_handler.f90 ─────────────────
!##############################################################
subroutine err_handle(err, err_code, called_from, file_name, &
    line_no, object_name, custom_1, custom_2, custom_3)
  ! Public: Stores a message in the error handler
  ! Meaning of err_code
  !     0 = no error
  ! 1-99: I/O errors
  !     1 = file could not be opened for read-access
  !     2 = file could not be opened for write-access
  !     3 = error in reading file
  !     4 = error in writing to file
  !     5 = error in reading file: EOR/EOF encountered
  !     6 = file open for write-access could not be closed
  !     7 = file open for read-access could not be closed
  ! 100-199: numerical errors
  !     100 = matrix apparently singular
  !     101 = matrix not positive definite
  !     102 = attempted division by zero
  !     103 = attempted logarithm of non-positive number
  !     104 = argument to exp function too large
  !     105 = attempted square root of negative number
  ! 200-299: memory errors
  !     200 = unable to dynamically allocate memory for object
  !     201 = unable to deallocate memory for object
  ! 300-399: array dimension errors
  !     300 = non-square matrix encountered where square
  !         matrix expected
  !     301 = dimensions of matrix arguments not conformable
  ! 1000: other error
  !   1000 = reserved for custom error messages
  implicit none
  ! declare required arguments
  type(error_type), intent(inout) :: err
  integer, intent(in) :: err_code
  ! declare optional arguments
  character (len=*), optional :: called_from, file_name, &
      object_name, custom_1, custom_2, custom_3
  integer, optional :: line_no
  ! local variables
```

```
      character(len=12) :: ichar
      ! begin
      select case(err_code)
         case(0)
            call insert_msg_line( &
                "No errors", err)
         ! I/O errors
         case(1)
            call insert_msg_line( &
                "File could not be opened for read-access", err)
         case(2)
            call insert_msg_line( &
                "File could not be opened for write-access", err)

         ! lines omitted for brevity

         case(301)
            call insert_msg_line( &
                "Dimensions of matrix arguments not conformable", &
                err)
         ! custom error message
         case(1000)
             ! don't do anything yet
         ! anything else
         case default
            call insert_msg_line("Unknown error code.", err)
      end select
      ! append other optional information if present
      if( present(custom_1) ) &
         call insert_msg_line(custom_1, err)
      if( present(custom_2) ) &
         call insert_msg_line(custom_2, err)
      if( present(custom_3) ) &
         call insert_msg_line(custom_3, err)
      if(present(file_name)) &
         call insert_msg_line("FILE: " // trim(file_name), err)
      if(present(line_no)) then
         write(ichar,"(I12)") line_no
         ichar = adjustl(ichar)
         call insert_msg_line("LINE: " // trim(ichar), err)
      end if
      if(present(object_name)) &
         call insert_msg_line(trim(object_name), err)
      if(present(called_from)) &
         call insert_msg_line("OCCURRED IN: " // &
             trim(called_from), err)
   end subroutine err_handle
   !################################################################
```

Finally, the public procedure `err_get_msgs` retrieves the stored messages from the object as a single string. In the string, message lines are separated by characters that signal the on-screen printing mechanism to advance to new lines. In Windows operating systems, a new line is indicated by a carriage return (CR) (character 13 in the ASCII collating sequence) followed by a line feed (LF) (ASCII character 10). Unix and Linux use only LF, whereas Macintosh® systems use CR. These nonprinting characters can be inserted through the Fortran intrinsic function achar.

```fortran
───────────── error_handler.f90 ─────────────
!####################################################################
subroutine err_get_msgs(err, msg_string, platform)
    ! Public: Retrieves all stored messages as a single character
    ! string, with message lines separated by platform-appropriate
    ! ASCII carriage control characters.
    ! Values for platform may be "UNIX", "MAC" or "PC"
    implicit none
    ! required arguments
    type(error_type), intent(inout) :: err
    character(len=*), intent(out) :: msg_string
    ! optional arguments
    character(len=*), intent(in), optional :: platform
    ! local variables
    character(len=4) :: plat
    integer :: posn
    logical :: first_time
    type(msg_line_type), pointer :: cur_line
    ! determine platform
    if( present(platform) ) then
        plat = platform
    else
        plat = "PC"
    end if
    ! clean out msg_string
    msg_string = ""
    ! step through the linked list, appending the lines
    first_time = .true.
    cur_line => err%head
    do
        if( .not.associated(cur_line) ) exit
        posn = len_trim(msg_string)
        if( (posn+3) >= len(msg_string) ) exit ! out of space
        posn = posn + 1
        if( .not.first_time) then
            select case(plat)
                case("UNIX")
                    ! Separate lines with LF
                    msg_string(posn:) = achar(10)
                    posn = posn + 1
                case("MAC")
                    ! Separate lines with CR
                    msg_string(posn:) = achar(13)
                    posn = posn + 1
                case default
                    msg_string(posn:) = achar(13) // achar(10)
                    posn = posn + 2
```

```
        end select
      end if
      msg_string(posn:) = trim(cur_line%line)
      first_time = .false.
      cur_line => cur_line%next
    end do
  end subroutine err_get_msgs
  !####################################################################
```

3.6.4   Using the Module

Notice that our module contains the definition of **error_type** and its associated procedures, but it does not contain any instances of the type. As discussed in Section 3.2, it's best to declare instances not within the module itself but in the program that uses the module. The instance can then be passed by argument to any procedure that may generate an error.

Example: In-Place Cholesky Factorization

If A is a symmetric, positive-definite matrix, then there is a unique lower-triangular matrix C, called the Cholesky factor, whose diagonal elements are all positive and that satisfies $CC^T = A$. Below is a module containing a procedure that overwrites the lower triangle of a matrix with its Cholesky factor based on an algorithm given by Thisted (1988). The dummy argument for the input matrix has assumed shape. The procedure will fail if the actual argument is not square or not positive.

```
──────────────────────────── cholesky.f90 ────────────────────────────
!####################################################################
module cholesky
  use error_handler
  implicit none
  private ! by default
  public :: cholesky_lower
  character(len=*), parameter :: modname = "cholesky"
contains
  integer function cholesky_lower(a, err) result(answer)
    !### Overwrites lower triangle of a symmetric, pos.-def. matrix
    !### with its cholesky factor. The upper triangle is untouched.
    !### Returns 0 if successful, 1 if failed.
    implicit none
    ! declare arguments
    double precision, intent(inout) :: a(:,:)
    type(error_type), intent(inout) :: err
    ! declare local variables and parameters
    character(len=*), parameter :: subname = "cholesky_lower"
    integer :: p, i, j, k
    double precision :: sum
    ! begin
    answer = 1
```

```
      p = size(a,1)
      if( p /= size(a,2) ) goto 700
      do j = 1, p
         sum = 0.D0
         do k = 1, j-1
            sum = sum + a(j,k)**2
         end do
         if( sum >= a(j,j) ) goto 710
         a(j,j) = sqrt( a(j,j) - sum )
         do i = j + 1, p
            sum = 0.D0
            do k = 1, j-1
               sum = sum + a(j,k)*a(i,k)
            end do
            if( a(j,j) == 0.D0 ) goto 710
            a(i,j) = ( a(i,j) - sum ) / a(j,j)
         end do
      end do
      ! normal exit
      answer = 0
      return
      ! error traps
700   call err_handle(err, 300, &
         called_from = subname//" in MOD "//modname)
      return
710   call err_handle(err, 101, &
         called_from = subname//" in MOD "//modname)
      return
   end function cholesky_lower
   !#################################################################
end module cholesky
!#################################################################
```

 Style tip————————————————————————————————————

Character-string parameters containing the module and procedure names
provide a convenient way to pass these names to the error handler.

___

Below is a simple program for testing the Cholesky routine on a 3×3
positive-definite matrix.

```
———————————————————————— err_test.f90 ————————
!#################################################################
!###    A simple program for testing the error_handler module    #######
!#################################################################
program err_test
   use error_handler
   use cholesky
   implicit none
   ! declare variables and parameters
   double precision :: s(3,3) = 0.D0
```

```
   integer :: i, j
   type(error_type) :: err
   character(len=*), parameter :: platform = "PC"
   character(len=256) :: msg_string
   ! enter a pos.def. matrix into lower triangle
   s(1,1) =  0.75
   s(2,1) = -0.20
   s(3,1) = -0.04
   s(2,2) =  4.66
   s(3,2) =  1.43
   s(3,3) =  5.45
   ! attempt cholesky factorization and print result
   if( cholesky_lower(s, err) /= 0 ) goto 800
   do i = 1, 3
      print "(3F10.5)", ( s(i,j), j=1, 3)
   end do
800 continue
   ! report error message
   if( err_msg_present(err) ) then
      call err_get_msgs(err, msg_string, platform)
      print "(A)", trim(msg_string)
      print "(A)", "Aborted"
   end if
end program err_test
!##############################################################################
```

The screen output from this test program looks like this:

```
d:\jls\software\demos>err_test
    0.86603    0.00000    0.00000
   -0.23094    2.14631    0.00000
   -0.04619    0.66129    2.23843
```

If the $(3, 1)$ element of the matrix is changed to

```
   s(3,1) = -2.00
```

then an error occurs:

```
d:\jls\software\demos>err_test
Matrix not positive definite
OCCURRED IN: cholesky_lower in MOD cholesky
Aborted
```

If `cholesky_in_place` were called from another procedure and an error occurred, then placing statements such as

```
   if( cholesky_lower(s, err) /= 0 ) goto 800
   ! lines omitted

   ! error trap
800 call err_handle(err, 1000, &
```

```
         called_from = subname//" in MOD "//modname)
      return
```

in the calling procedure would append the name of that procedure to the
error message list for traceback purposes.

3.6.5   General Guidelines for Modules

We conclude this chapter with some general guidelines for creating modules
for pseudo object-oriented programming in Fortran.

1. A module should contain no variables, whether public or private.
   Local variables within module *procedures* are acceptable, but there
   should be no variables declared in a module before the `contains`
   statement.

2. Module parameters are acceptable, both public and private. But consider carefully before making a parameter public because from that
   point on the parameter's name is effectively retired and cannot be
   used for any other purpose within program units that use the module.

3. Placing derived-type definitions within modules is encouraged. These
   derived types will often be made public, but in most cases it's a good
   idea to keep the contents (i.e., components) of the types private.

4. If a module contains a public derived type whose contents are private,
   then communication between an instance of the type and the outside
   should be accomplished by public functions or subroutines placed in
   the same module.

5. When developing a module, do not presume that your public functions or subroutines will be used intelligently or correctly by other
   program units—even if the developer of the other program units is
   you. Rather, assume that the procedures will be misused. Extensively
   check the input arguments for correctness and consistency, and provide some means for reporting inconsistencies and error messages to
   the outside world (e.g., by using the `error_handler` module).

6. Make your source code self-documenting by including extensive comments, especially in public functions and subroutines. This small in-
   vestment of time and effort will greatly pay off in the future.

Don't fall into the trap of thinking that if you write a module today,
you will remember the details of the module's inner workings in the weeks,
months, or years ahead. Don't presume that you will even remember how
to use the module's public procedures correctly. In the future, you will be
a different person from the one you are today. Therefore, strive to make

each module as self-contained as possible so that it acts as an intelligent "black box." Strive to make the module crash-proof by including extensive checks within the public procedures.

3.7 Additional Exercises

1. The modified Gram-Schmidt procedure for orthogonalizing a matrix was presented in Exercise 7 of Section 2.6. Implement this method in a function or subroutine that uses the error handler module.

2. Our error handler module can be enhanced in many ways. In the `err_handle` procedure, for example, one could add an optional argument **append** that, depending on whether it is **.true.** or **.false.**, appends the new error message lines to the current list or wipes out the stored messages to start a new list. Revise `error_handler` to add this new feature and any other features that you may find useful, but do it in such a way that existing applications that may already use the module will not need to be revised.

3. The parameters of a multivariate normal distribution are usually expressed as $\theta = (\mu, \Sigma)$, where μ is a mean vector of length p and Σ is a $p \times p$ covariance matrix. Create a derived type for storing θ as a single object, and place your type definition in a module. Write a module procedure for computing the maximum-likelihood estimate of θ from an $n \times p$ data matrix.

4. Many datasets can be represented in a rectangular form, with rows corresponding to observational units and columns corresponding to variables. If all of the variables are numeric, then the dataset can be stored very simply as a rank-two real array. If both numeric and nonnumeric (e.g., nominal) variables are present, however, it may be better to store the dataset in some other fashion. Devise a strategy for storing variables in a linked list, with one node per variable, and implement it in a Fortran module.

5. The empirical distribution function $\hat{F}(x)$, which assigns probability mass $1/n$ to each of the data values $x_1, x_2, \ldots, x_n$ in a sample, plays a central role in the bootstrap (Efron and Tibshirani, 1994) and other nonparametric statistical techniques. Create a derived type for storing an empirical distribution function, with procedures for computing it from a sample and for drawing random samples from it. For computing $\hat{F}(x)$, you may find it helpful to sort the data values with a binary tree, as discussed in Exercises 7 and 8 of Section 3.5.5.

4

Implementing Computational Routines

The last chapter introduced the key ideas of modules, derived types and pointers in Fortran, showing how they form the basis of a new pseudo object-oriented style. We now turn our attention to the heart of a statistical application, the computational routines. Efficient computation has always been one of the most appealing aspects of Fortran. Volumes of procedures written since the late 1970s, many of them in FORTRAN 77, are still widely used. In this chapter, we discuss how to incorporate new features of the Fortran language into computational routines without sacrificing good performance.

4.1   Issues of Numerical Accuracy

4.1.1   Accuracy Is Crucial

First and foremost, a computational routine should produce accurate results. Accuracy should not be sacrificed to enhance speed, to conserve memory, or for any other reason. In the early days of computing, many routines developed by scientists and engineers were seriously flawed. For example, the infamous pseudorandom generator RANDU developed at IBM and used by many throughout the 1970s was later found to be "really horrible" (Knuth, 1981, p. 173). Fortunately, we have learned much from past mistakes. Many texts on numerical methods are available that are filled with cautionary tales to help us avoid dangerous pitfalls. Golub and van Loan (1996) provide excellent coverage of general matrix computations, whereas Kennedy and Gentle (1980), Thisted (1988), and Lange (1999) write specifically for a statistical audience. A respectable treatment of numerical accuracy and error analysis is beyond the scope of this book. We would feel irresponsible, however, if we did not mention some key ideas and provide a few simple guidelines dictated by common sense.

4.1.2   Floating-Point Approximation

When writing computational routines, the first thing to realize is that non-integer computer arithmetic is inexact. Real numbers are rationally approximated by floating-point models that may vary across platforms. With any given compiler, different models are made available by changing the kind parameter as described in Section 2.1.7.

A typical 64-bit real number, which corresponds to double precision, is precise to 15 significant digits and has a decimal exponent range of 307; it takes a maximum value of

$$1.797693134862316 \times 10^{308} \approx e^{709.7827}$$

and a minimum positive value of

$$2.225073858507201 \times 10^{-308} \approx e^{-708.3964}.$$

In this model, the number $1 + \epsilon$ for any

$$0 < \epsilon < 2.220446049250313 \times 10^{-16}$$

will be recognized as 1. In contrast, a typical 32-bit or single-precision real number has only 6 significant digits and takes a maximum value of 3.4028235×10^{38} and a minimum positive value of $1.1754944 \times 10^{-38}$.

The first step in maintaining accuracy is to make sure that you understand what kind of real variables you are using. Simply declaring

```
real :: x
```

is not good enough; the kind parameter should be explicitly set. This can be confusing, as the values of kind corresponding to the different floating-point models vary from one compiler to another. Consult your compiler's documentation to learn what kinds are available. Information is also available through several Fortran intrinsic query functions. If x is a real variable of any kind, then precision(x) returns the number of significant digits of decimal precision retained by this kind; range(x) returns the range of decimal exponents that can be represented; huge(x) returns the largest representable value; and tiny(x) returns the smallest positive value. As an experiment, run this test program to see the characteristics of your compiler's default real kind.

```
program test
  real :: x
  print *, precision(x)
  print *, range(x)
  print *, huge(x)
  print *, tiny(x)
end program test
```

In the days when physical memory was scarce, programmers had to worry about whether to store values in single or double precision and make painful decisions about accuracy versus the ability to handle large datasets. Today, unless we are processing large volumes of data, it often makes sense simply to declare all real variables to have 64 bits. As the kind parameters corresponding to 64-bit reals may vary, it is convenient to select the kind automatically using the Fortran intrinsic function selected_real_kind. The declarations

```
integer, parameter :: my_dble = selected_real_kind(15,307)
real(kind=my_dble) :: x
```

automatically define x to be the smallest real kind with at least 15 significant digits of precision and a decimal exponent range of at least 307. A similar intrinsic function is available for integers. The declaration

```
integer, parameter :: my_int = selected_int_kind(9)
```

identifies the shortest kind capable of holding numbers between -10^9 and 10^9. That is, it will select a 32-bit integer, which has a maximum value of $2^{31} - 1 = 2{,}147{,}483{,}647$. The functions range and huge can also be applied to integer variables.

4.1.3   Roundoff and Cancellation Error

Reasonable measures should be taken to prevent large errors of cancellation. Despite the algebraic identity

$$\sum_{i=1}^{n}(x_i - \bar{x})^2 = \sum_{i=1}^{n}x_i^2 - n\bar{x}^2, \qquad (4.1)$$

it is well known that the errors produced by the method on the right-hand side can be disastrous. The problem is that when the dataset has a small coefficient of variation—i.e., when the sample mean $\bar{x} = n^{-1}\sum_i x_i$ is large relative to the standard deviation—then $\sum_i x_i^2$ and $n\bar{x}^2$ become relatively close, differing only in their final digits. The roundoff errors that arise in the accumulation of $\sum_i x_i$ and $\sum_i x_i^2$ then move to the leading digits when the difference is taken. When computing sample variances and covariances, it's safer to pass through the data twice—first to compute $\bar{x}$

and then to accumulate the sum of $(x_i - \bar{x})^2$—even though it takes more time. Alternatively, you can use the one-pass updating method described in Exercise 10 of Section 2.6. Another famous example of where catastrophic cancellation error may occur is in the computation of e^x for $x < 0$ by Taylor series. A good introduction to the causes and impact of roundoff error in scientific computation is given by Kahaner, Moler, and Nash (1988).

To reduce the effects of roundoff error, sums of real numbers should always be accumulated in double precision. The product of two single-precision reals can be computed and stored exactly in double precision. Therefore, if we compute the inner product of two single-precision real vectors in double precision, no precision is lost. Large errors of rounding and truncation may be inadvertently introduced through mixed-type expressions involving integers and reals, as described in Section 2.5.2, so it's a good idea to avoid mixed-type expressions. Also avoid integer division unless you are absolutely sure that you know what you are doing.

Because floating-point arithmetic is approximate, it does not have many of the familiar properties of ordinary arithmetic. Floating-point addition is not associative, so $(x + y) + z$ may be slightly different from $x + (y + z)$. When computing with matrices, a matrix result that is theoretically symmetric may not be exactly symmetric. If exact symmetry is required, we can replace a(i,j) and a(j,i) by their average for all i > j. Or we can compute the upper triangle of the matrix and copy its contents into the lower triangle, eliminating calculations that are redundant.

Some compilers offer integers longer than 32 bits or floating-point models with precision and range beyond 64-bit reals. These extended-range representations may be helpful in certain circumstances but should not be used as a matter of routine, as they may cause a program to be slow and nonportable.

4.1.4   Arithmetic Exceptions

A computational result that exceeds the largest possible value for a variable of that kind is called an overflow. The effect of an overflow may vary from one compiler to another, and it may also vary according to the options chosen when the code is compiled. In some cases, an overflow will cause the program to immediately terminate. In other cases, the result may be represented by a special code such as NaN or Inf, and the program may continue to run. If execution continues, these undefined values are likely to propagate to other program variables and eventually cause a crash.

Because measurements like 10^{307} are uncommon in the real world, one might think that overflows in 64-bit reals are so rare that we needn't worry about them. This is simply not true. Numbers of this magnitude do arise as intermediate quantities in computations whose final results are reasonable. For example, consider the density function for a Student's t random variable with $\nu > 0$ degrees of freedom,

$$f(y) = \frac{\Gamma\left(\frac{\nu+1}{2}\right)}{\Gamma\left(\frac{\nu}{2}\right)\sqrt{\pi\nu}} \left(1 + \frac{y^2}{\nu}\right)^{-\left(\frac{\nu+1}{2}\right)}. \tag{4.2}$$

The gamma function in the numerator produces a double-precision overflow for $\nu = 343$. A simple remedy in this case is to work with the log of the gamma function.

Many compilers offer special platform-specific tools for handling overflows. If you are reluctant to invest the time and effort to learn how to use these tools, or if you want to maintain portability, a good strategy is to anticipate where overflows are likely to occur and prevent them from happening. This is not as difficult as it may seem. With 64-bit real variables, overflows caused by summation are rare. The main culprit is exponentiation. Before evaluating any expression involving exp(x), for example, one may first compare x with log(huge(x)); if x is too large, we can report the problem to our own error-messaging system (Section 3.6) and gracefully exit. Another common culprit is cumulative multiplication, which arises in computing likelihood functions and determinants of large matrices. To calculate $\prod_i x_i$, a less efficient but safer method is to accumulate $\sum_i \log|x_i|$.

Any real result of magnitude less than the smallest representable value produces an underflow . When an underflow occurs, the usual default behavior is to set the result to zero and continue. In most cases, this is acceptable. We should, however, make sure that a zero value does not cause any problems later (e.g., through attempted division by zero). As a matter of routine, one should incorporate checks to prevent division by zero, logarithms of nonpositive numbers, square roots of negative numbers, and so on.

Many novice programmers fail to put enough safeguards into their computational routines, and the resulting programs crash too often. Carried to an extreme, however, too much vigilance may cause the programmer and the program itself to get bogged down. With enough safeguards, it is theoretically possible to thwart every conceivable arithmetic exception that might ever occur, but the code would be too cumbersome to write and would run too slowly. Expert programmers who produce very-high-performance numerical procedures, such as those in LAPACK, strike a careful balance between preventing exceptions through coded safeguards and allowing exceptions to occur, with the subsequent actions determined by special exception-handling procedures. A useful discussion on this balance is given by Hauser (1996). For most of us, the commonsense strategy is to add enough safeguards to prevent crashes during routine use and in exceptional circumstances, yet allow them to be theoretically possible under rare and highly extreme conditions.

4.1.5 Resources for Numerical Programming

Many of the computational tasks required within a statistical application—solving linear systems, calculating eigenvalues and eigenvectors, computing determinants, finding tail areas of probability distributions, etc.—can be done in a variety of ways. Some of these tend to be numerically stable and accurate, whereas others are inherently ill-conditioned and susceptible to major error. One excellent way to ensure accuracy and efficiency is to call external procedures that have been professionally coded and tested. Many high-performance procedures are available in LAPACK or libraries from the Numerical Algorithms Group (NAG) or IMSL. If the source code is provided, the source can simply be copied and placed in your own program. If the source is not available, you will have to link to the procedures either during the build process or at run time using dynamic link libraries (DLLs). If you choose to use procedures from any of these libraries, make sure that you understand the implications of any copyright restrictions or licensing requirements before distributing your program to others. On the other hand, if you choose to code a procedure yourself, take some time to learn about the properties of the algorithm that you intend to use.

The well-known *Numerical Recipes* textbooks by Press et al. (1992, 1996) are invaluable references for scientific programming in Fortran, that contain many open-source procedures for performing matrix computations, approximating functions, numerical integration, and so on. Until 1997, hundreds of procedures of interest to statisticians were published in *Applied Statistics*, and they have been partially collected by Griffiths and Hill (1985). Most of these procedures were written in FORTRAN 77, and some of them will need revision to conform to practices suggested in this book—e.g., changing from single to double precision and eliminating the use of pause and stop. Adaptations of some *Applied Statistics* algorithms to modern Fortran are publicly available from the Statlib archives at Carnegie Mellon University and other Web sites. Source code from the Internet may be useful, but it comes with no guarantees of accuracy or good performance; be sure to test it thoroughly and give proper credit to the authors if you find it useful.

4.1.6 Exercises

1. Write a procedure for computing a sample variance with single precision using the right-hand side of the identity (4.1), and show by example how the result can be quite inaccurate.

2. Write a sample program to demonstrate that floating-point addition is not associative.

3. The Taylor series

$$e^x = 1 + x + \frac{x^2}{2!} + \frac{x^2}{3!} + \cdots$$

converges for all values of x. This suggests that $y = e^x$ may be computed as follows: set $y = a = n = 1$ and repeat

$$
\begin{aligned}
a &= ax/n, \\
y &= y + a, \\
n &= n + 1,
\end{aligned}
$$

until a is indistinguishable from zero. Implement this procedure in a Fortran procedure using single-precision arithmetic. Compare your computed values of e^x to those from the intrinsic function exp for increasingly negative values $x = -1, -2, -3, \dots$ until exp(x) underflows. Try to explain what you find. To solve the problem, modify your function to compute $e^{|x|}$ and take its reciprocal if $x < 0$.

4. There are many ways to compute tail areas from a normal distribution,

$$
1 - \Phi(z) = \frac{1}{\sqrt{2\pi}} \int_z^\infty e^{-t^2/2} dt.
$$

One way uses a continued fraction derived by Laplace (1749–1827),

$$
\int_z^\infty e^{-t^2/2} dt = \cfrac{e^{-z^2/2}}{z + \cfrac{1}{z + \cfrac{2}{z + \cfrac{3}{z + \cdots}}}}.
$$

Write a procedure for approximating $\Phi(z)$ to any given number of terms in the continued fraction. Compare your results with values for $\Phi(z)$ obtained from published tables or from one or more standard statistical packages.

5. To compute the gamma function

$$
\Gamma(x) = \int_0^\infty y^{x-1} e^{-y} dy
$$

for $x > 0$, Press et al. (1992) recommend the approximation

$$
\begin{aligned}
\Gamma(x) \approx\ & x^{-1} \left(x + \frac{11}{2} \right)^{x+1/2} e^{-(x+11/2)} \\
& \times \sqrt{2\pi} \left[c_0 + \frac{c_1}{x+1} + \frac{c_2}{x+2} + \cdots + \frac{c_6}{x+6} \right],
\end{aligned}
$$

with coefficients as shown in Table 4.1. Using this method, write a procedure for evaluating $\log \Gamma(x)$ in double precision. Investigate its accuracy by comparing your estimates of $\Gamma(x+1)$ with exact values of $x!$ computed by integer arithmetic for $x = 0, 1, 2, \dots$.

TABLE 4.1. Coefficients for approximating the gamma function.

c_0	1.000000000190015
c_1	76.18009172947146
c_2	-86.50532032941677
c_3	24.01409824083091
c_4	-1.231739572450155
c_5	$1.208650973866179 \times 10^{-3}$
c_6	$-5.395239384953 \times 10^{-6}$

6. Using your log-gamma approximation from the previous exercise, write a procedure for computing the Student's t density function (4.2). Compare your results to those from a well-established statistical package for reasonable values of y and ν.

7. Many interesting and elegant formulas exist for the constant π, but they are not equally good for computation. For example, Wallis (1616–1703) found that

$$\frac{\pi}{2} = \frac{2 \cdot 2 \cdot 4 \cdot 4 \cdot 6 \cdot 6 \cdot 8 \cdot 8 \cdots}{1 \cdot 3 \cdot 3 \cdot 5 \cdot 5 \cdot 7 \cdot 7 \cdot 9 \cdots}$$

and

$$\frac{2}{\pi} = \left(1 - \frac{1}{2^2}\right)\left(1 - \frac{1}{4^2}\right)\left(1 - \frac{1}{6^2}\right)\cdots.$$

A continued fraction due to Brouncker (1620–1684) is

$$\frac{4}{\pi} = 1 + \cfrac{1^2}{2 + \cfrac{3^2}{2 + \cfrac{5^2}{2 + \cfrac{7^2}{2 + \cdots}}}}.$$

Euler (1707–1783) presented the well-known series

$$\frac{\pi^2}{6} = \frac{1}{1^2} + \frac{1}{2^2} + \frac{1}{3^2} + \frac{1}{4^2} + \cdots,$$

and an early product derived by Vieta (1540–1603) is

$$\frac{2}{\pi} = \frac{\sqrt{2}}{2} \cdot \frac{\sqrt{2 + \sqrt{2}}}{2} \cdot \frac{\sqrt{2 + \sqrt{2 + \sqrt{2}}}}{2} \cdots.$$

Implement each of these methods in double-precision arithmetic and see how many terms you need to match $\pi \approx 3.14159265$ to single-precision accuracy.

4.2 Example: Fitting a Simple Finite Mixture

4.2.1 The Problem

To illustrate some good practices for designing and coding computational routines, we now turn to a simple example. Suppose we have a sample $y_1, \ldots, y_n$ from a population with density

$$f(y_i) = \pi \lambda_1 \exp(-\lambda_1 y_i) + (1 - \pi) \lambda_2 \exp(-\lambda_2 y_i),$$

$y_i > 0$, where the parameters $\pi \in [0, 1]$, $\lambda_1 > 0$, and $\lambda_2 > 0$ are to be estimated. That is, some proportion π of the population follows an exponential distribution with mean λ_1^{-1}, and the remainder follows an exponential distribution with mean λ_2^{-1}. This is one of the simplest examples of a finite-mixture model (McLachlan and Peel, 2000). Maximum-likelihood (ML) estimates can be easily calculated by the following EM algorithm. First, calculate the posterior probability that y_i came from the first population component,

$$\delta_i = \frac{\pi \lambda_1 \exp(-\lambda_1 y_i)}{\pi \lambda_1 \exp(-\lambda_1 y_i) + (1 - \pi) \lambda_2 \exp(-\lambda_2 y_i)},$$

$i = 1, \ldots, n$, using provisional estimates $\pi = \hat{\pi}$, $\lambda_1 = \hat{\lambda}_1$ and $\lambda_2 = \hat{\lambda}_2$. Second, update the estimates by

$$\hat{\pi} = \frac{\sum_i \delta_i}{n}, \quad \hat{\lambda}_1 = \frac{\sum_i \delta_i}{\sum_i \delta_i y_i}, \quad \hat{\lambda}_2 = \frac{\sum_i (1 - \delta_i)}{\sum_i (1 - \delta_i) y_i}.$$

Alternating between these two steps produces a sequence of estimates that converges reliably to a local or global maximum of the loglikelihood function $l = \sum_i \log f(y_i)$. If the sample is large enough for the loglikelihood function to be approximately quadratic in the vicinity of the ML estimate, then reasonably accurate inferences may be obtained from the estimated covariance matrix $[-l'']^{-1}$, where l'' is the Hessian (i.e., the matrix of second derivatives of l with respect to the parameters). For this model, the score functions are

$$\frac{\partial l}{\partial \pi} = \frac{\sum_i (\delta_i - \pi)}{\pi (1 - \pi)},$$

$$\frac{\partial l}{\partial \lambda_1} = -\sum_i \delta_i (y_i - \lambda_1^{-1}),$$

$$\frac{\partial l}{\partial \lambda_2} = -\sum_i (1 - \delta_i)(y_i - \lambda_2^{-1}),$$

the diagonal elements of the Hessian are

$$\frac{\partial^2 l}{\partial \pi^2} = -\frac{\sum_i (\delta_i - \pi)^2}{\pi^2 (1 - \pi)^2},$$

$$\frac{\partial^2 l}{\partial \lambda_1^2} = \sum_i \left\{ -\delta_i \lambda_1^{-2} + \delta_i(1 - \delta_i)(y_i - \lambda_1^{-1})^2 \right\},$$

$$\frac{\partial^2 l}{\partial \lambda_2^2} = \sum_i \left\{ -(1 - \delta_i)\lambda_2^{-2} + \delta_i(1 - \delta_i)(y_i - \lambda_2^{-1})^2 \right\},$$

and the off-diagonal elements are

$$\frac{\partial^2 l}{\partial \pi \partial \lambda_1} = -\frac{\sum_i \delta_i(1 - \delta_i)(y_i - \lambda_1^{-1})}{\pi(1 - \pi)},$$

$$\frac{\partial^2 l}{\partial \pi \partial \lambda_2} = \frac{\sum_i \delta_i(1 - \delta_i)(y_i - \lambda_2^{-1})}{\pi(1 - \pi)},$$

and

$$\frac{\partial^2 l}{\partial \lambda_1 \partial \lambda_2} = -\sum_i \delta_i(1 - \delta_i)(y_i - \lambda_1^{-1})(y_i - \lambda_2^{-1}).$$

Likelihood functions for finite mixtures have unusual properties that can make statistical inference challenging (Titterington, Smith, and Makov, 1985; McLachlan and Peel, 2000). Difficulties in likelihood-based inference for this model were investigated by Chung, Loken, and Schafer (2004). Nevertheless, ML via the EM algorithm remains the most widely used estimation technique for finite mixtures. EM increases the loglikelihood at each step and will eventually converge to something, but the solution may not be well-behaved. Sometimes EM will converge to an estimate on the boundary of the parameter space. Likelihood functions for finite mixtures are invariant to permutations of the component labels. In this problem, if a mode occurs at $\pi = 0.4$, $\lambda_1 = 1.0$, $\lambda_2 = 3.0$, then an equivalent mode will exist at $\pi = 0.6$, $\lambda_1 = 3.0$, $\lambda_2 = 1.0$, and EM could converge to either one depending on the starting values. The parameter space also has continuous regions of indeterminacy. In this example, π is indeterminate where $\lambda_1 = \lambda_2$, and the Hessian matrix is singular anywhere within this plane. For these reasons, it is a good idea to examine the score functions at the EM solution to see whether they are zero and to check the Hessian to see if the loglikelihood is concave.

4.2.2 Programming Constants

We will now write a Fortran procedure that implements the EM algorithm and computes the first and second derivatives at the solution. Our first act of coding is to create a module of programming constants. This module has no executable statements but merely provides a convenient location to define parameters that will be used throughout the program. In particular, this is where we will define the kind parameters for integer and real variables to help ensure consistency of results across compilers, as discussed in Section 4.1.2. A bare-bones module of constants is shown below.

```
———————————— constants.f90 ————————————
!##################################################################
module program_constants
   implicit none
   public
   ! Define compiler-specific KIND numbers for integers,
   ! single and double-precision reals to help ensure consistency
   ! of performance across platforms:
   integer, parameter :: our_int = selected_int_kind(9), &
       our_sgle = selected_real_kind(6,37), &
       our_dble = selected_real_kind(15,307)
   ! Common integer values returned by all functions to indicate
   ! success or failure:
   integer(kind=our_int), parameter :: RETURN_SUCCESS = 0, &
       RETURN_FAIL = -1
end module program_constants
8!################################################################
```

 Style tip————————————————————————————————

Place important parameters in a module so that, if it becomes necessary
to change them, the changes can be made quickly.

4.2.3   A Computational Engine Module

When creating a computational routine, whether it will be used for a single
application or shared among applications, we highly recommend placing it
in a module. Making it a public procedure within a module has many benefits. First, the procedure will automatically be given an explicit interface.
As a result, during the build process, the linker will check to make sure
that when a program calls the procedure, the actual and dummy arguments all agree in type and kind. With a module procedure, the argument
list may use enhanced features such as assumed-shape arrays and optional
arguments to make the calling process more convenient and less prone to
error. Most importantly, if the procedure is placed within a module, all of
its inner workings—the parameters, data types, and procedures used by
it—can be kept private, so that future changes and enhancements are less
likely to propagate errors.

Here is the skeleton of the module that will hold the computational engine. This module has one public procedure; everything else is kept private.

```
———————————— em_exponential_engine.f90 ————————————
!##################################################################
module em_exponential_engine
   use error_handler
   use program_constants
```

```
    implicit none
    private ! by default
    public :: run_em_exponential
    ! parameters private to this module
    character(len=*), parameter :: modname = "em_exponential_engine"
    !###########################################################
contains
    !###########################################################
    ! all public and private procedures will be placed here
    !###########################################################
end module em_exponential_engine
!###########################################################
```

4.2.4 A Public Procedure With Safeguards

Now we begin to write the procedure itself. Because it is public, we should
presume that it will be misused and put in safeguards to prevent common
errors and provide feedback through informative messages. Here are the
parts of the procedure that define its interface with the outside world.

```
    !###############################################################
    integer(kind=our_int) function run_em_exponential( y, &
        pi, lambda_1, lambda_2, iter, converged, loglik, &
        score, hessian, err, maxits, eps) result(answer)
    ! EM algorithm for computing ML estimates for the mixture of
    ! two exponentials,
    !           pi * exponential with mean 1/lambda_1
    !    + (1 - pi) * exponential with mean 1/lambda_2
    implicit none
    !  Input data containing the sample:
    real(kind=our_dble), intent(in) :: y(:)
    !  Starting values for parameters; these will also return
    !  the estimates:
    real(kind=our_dble), intent(inout) :: pi, lambda_1, lambda_2
    !  Number of EM iterations performed:
    integer(kind=our_int), intent(out) :: iter
    !  T if EM converged, F otherwise:
    logical, intent(out) :: converged
    !  Loglikelihood function and its first two derivatives at
    !  the parameter estimates:
    real(kind=our_dble), intent(out) :: loglik, score(3), &
        hessian(3,3)
    !  Error messages:
    type(error_type), intent(inout) :: err
    !  Optional: maximum number of iterations and
    !  criterion for judging convergence.
    integer(kind=our_int), intent(in), optional :: maxits
    real(kind=our_dble), intent(in), optional :: eps
    ! locals
    integer(kind=our_int) :: max_iter, i, n
    real(kind=our_dble) :: epsilon, oldpi, oldlambda_1, &
```

```
            oldlambda_2, d, f, sumd, sumy, sumdy
      character(len=12) :: sInt
      character(len=*), parameter :: subname = "run_em_exponential"
      ! begin
      answer = RETURN_FAIL
      score(:) = 0.
      hessian(:,:) = 0.
      ! check input arguments and set defaults
      if( (pi < 0.) .or. (pi > 1.) ) goto 300
      if( ( lambda_1 <= 0. ) .or. ( lambda_2 <= 0. ) ) goto 400
      if( present(maxits) ) then
         if(maxits < 0) goto 500
         max_iter = maxits
      else
         max_iter = 1000
      end if
      if( present(eps) ) then
         if( eps < 0. ) goto 600
         epsilon = eps
      else
         epsilon = .00001
      end if
      ! run EM

      !!!! lines omitted !!!!

      ! normal exit
      answer = RETURN_SUCCESS
      return
      ! error traps
300   call err_handle(err, 1000, &
           called_from = subname//" in MOD "//modname, &
           custom_1 = "Argument pi out of range.")
      return
400   call err_handle(err, 1000, &
           called_from = subname//" in MOD "//modname, &
           custom_1 = "Argument lambda_1 or lambda_2 out of range.")
      return
500   call err_handle(err, 1000, &
           called_from = subname//" in MOD "//modname, &
           custom_1 = "Invalid value for argument maxits.")
      return
600   call err_handle(err, 1000, &
           called_from = subname//" in MOD "//modname, &
           custom_1 = "Invalid value for argument eps.")
      return

      !!!! lines omitted !!!!

   end function run_em_exponential
   !###############################################################
```

Notice how easy it is to add safeguards and error messages using the system that we developed in Section 3.6. If necessary, more can be added in the future. For example, we should probably check to make sure that all the data values in y are positive. But life is hectic, and despite our good intentions we may never actually do it later. Try to anticipate as many errors as you can and incorporate checks when writing the procedure for the first time. Notice also that we've added comments to describe the procedure and each of the dummy arguments. With these comments, an intelligent person won't need an extra document to understand how to use it. As a rule, force yourself to add comments generously to all public procedures so that they become self-documenting.

4.2.5   The Computations

The remaining parts of the function that implements the EM algorithm are shown below. The function's returned value is RETURN_SUCCESS if the algorithm runs normally, whether or not it has converged within the specified maximum number of iterations; a value of RETURN_FAIL indicates that execution had to be aborted to prevent a crash. This function calls a private procedure, eval_score_and_hessian, which is not shown; it is a straightforward implementation of derivative formulas given in Section 4.2.1.

```
! run EM
n = size(y)
if( n == 0 )  goto 700
sumy = sum(y)
converged = .false.
iter = 0
do
    iter = iter + 1
    write( sInt, "(I12)" ) iter
    sInt = adjustl( sInt )
    oldpi = pi
    oldlambda_1 = lambda_1
    oldlambda_2 = lambda_2
    loglik = 0.
    sumd = 0.
    sumdy = 0.
    do i = 1, n
        f = pi * lambda_1 * exp( -lambda_1 * y(i) ) &
            + (1 - pi) * lambda_2 * exp( -lambda_2 * y(i) )
        loglik = loglik + log(f)
        d = pi * lambda_1 * exp( -lambda_1 * y(i) ) / f
        sumd = sumd + d
        sumdy = sumdy + d * y(i)
    end do
    pi = sumd / real(n)
    if( sumdy == 0. ) goto 700
    lambda_1 = sumd / sumdy
```

```
           if( sumy == sumdy ) goto 700
           lambda_2 = ( real(n) - sumd ) / ( sumy - sumdy )
           converged = ( abs(pi - oldpi) < epsilon ) .and. &
                ( abs(lambda_1 - oldlambda_1 ) < epsilon ) .and. &
                ( abs(lambda_2 - oldlambda_2 ) < epsilon )
           if( ( iter >= max_iter ) .or. converged ) exit
        end do
        if( .not. converged ) &
            call err_handle(err, 1000, &
            called_from = subname//" in MOD "//modname, &
            custom_1 = "Algorithm failed to converge by " &
            // "iteration " // trim(sInt) )
        if( eval_score_and_hessian( y, pi, lambda_1, lambda_2, &
            score, hessian, err ) == RETURN_FAIL ) goto 800
        ! normal exit
        answer = RETURN_SUCCESS

        !!!! lines omitted !!!!

700     call err_handle(err, 1000, &
            called_from = subname//" in MOD "//modname, &
            custom_1 = "Attempted division by zero;", &
            custom_2 = "EM algorithm aborted at iteration " &
            // trim(sInt) )
        return
800     call err_handle(err, 1000, &
            called_from = subname//" in MOD "//modname)
        return
     end function run_em_exponential
    !#################################################################
```

 Style tip ──

If an iterative procedure must be aborted, report to the error handler the
iteration at which the problem occurred.

───

   This procedure as shown is still not entirely safe. For greater safety, we
should probably check to prevent overflows before the calls to exp, and we
should check for underflow in the variable f before computing log(f).

4.2.6 Strategies for Calling the Engine

The final, crucial step is to actually call the engine, test it, and use it. We
may approach this task in at least three ways. First, we can write a console
program that reads in a dataset, calls the EM procedure, and reports the
results. A second way is to encapsulate the procedure in a dynamic link
library (DLL) and call it from another application (e.g., a computational

package such as S-PLUS, R, or MATLAB). A third way is to package it as a COM server and call it from another application.

Each method has its own advantages and drawbacks. A stand-alone console program is highly portable. Writing a program for your own personal use that works on a single dataset is easy. But creating a robust application that can be used by others for a variety of datasets is not trivial; lots of code is needed to set up the user interface, to handle file I/O for reading the data, and to format, print, and save the results. Moreover, console applications tend to be poor environments for testing. Our program may produce answers, but we cannot be certain that the answers are correct until we carefully analyze them and obtain independent verification. This means that we may have to read the results into a computational package and do some sophisticated postprocessing; or we may need to try out the program on artificial, simulated, or published data for which the correct answers are already known. Either way, this requires sending data back and forth between our console program and a computational package, which can be tedious.

If we want to test the procedure as quickly as possible and then move on to other tasks, a better method is to produce a DLL. Most Fortran compilers can create DLLs quite easily. However, the methods by which applications communicate with DLLs may be crude or poorly documented. It is easy to make mistakes by passing data incorrectly, and if there is a problem, the system may not tell us where it occurred. With DLLs it may also be difficult to locate certain types of errors within the Fortran code (e.g., subscripts that stray outside the boundaries of an array). The process of creating and using DLLs will be described in Chapter 6.

The third option, creating a COM server, opens up elegant and robust channels of communication between your engine and the outside world. Ultimately, this may be the best way to package your software to make it versatile and easy for others to use. It does, however, require a fair amount of additional coding to create and document all the properties and methods. It also requires us to wrap our Fortran with additional, compiler-specific code to create the server. The subject of COM servers will be taken up in detail beginning with Chapter 7. Most statisticians will find these techniques useful for packaging and delivering the finished product but not for preliminary testing or incremental development.

4.2.7 A Simple Calling Program

As an initial test of our procedure, let's run the EM algorithm on this example dataset of $n = 50$ observations:

```
5.6   0.7   2.4   2.2   4.5   0.6   2.3   3.1   1.6   2.2
0.1   4.9   9.0   7.4   1.8   9.7   0.9   1.0   0.7   3.4
1.8   0.5   0.1   0.7   0.1   6.6   1.6   8.6   0.3   0.1
```

```
4.2  0.8  3.1  0.2  1.0  2.0  2.3  0.8  6.6  1.2
0.3  2.7  0.5  0.7  1.8  1.5  2.8 18.3  1.2  0.6
```

These data, exactly as shown, were placed in an ASCII text file called
good50.dat. Below is a crude test program that reads in this particular
dataset, calls the EM procedure with arbitrarily chosen starting values of
$\pi = 0.3$, $\lambda_1 = 0.5$, and $\lambda_2 = 2.0$, and prints all results to the console.

```fortran
———————————————————— em_test.f90 ————
!####################################################################
program em_test
   ! quick and dirty test program
   use error_handler
   use program_constants
   use em_exponential_engine
   implicit none
   ! declare variables
   type(error_type) :: err
   real(kind=our_dble) :: y(50), pi=.30, lambda_1=0.5, lambda_2=2.0, &
        loglik, score(3), hessian(3,3)
   integer(kind=our_int) :: iter, i
   logical :: converged
   character(len=*), parameter :: platform = "PC"
   character(len=256) :: msg_string
   ! begin
   open( 10, file="good50.dat" )
   read( 10, * ) ( y(i), i = 1, 50 )
   close( 10 )
   if( run_em_exponential( y, pi, lambda_1, lambda_2, iter, &
        converged, loglik, score, hessian, err ) == RETURN_FAIL ) &
        goto 800
   print *, "Iterations:", iter
   print *, "Converged:", converged
   print *, "Pi = ", pi
   print *, "Lambda_1 = ", lambda_1
   print *, "Lambda_2 = ", lambda_2
   print *, "Loglik = ", loglik
   print *, "Score:"
   print *, score(:)
   print *, "Hessian:"
   do i = 1, 3
      print *, hessian(i,:)
   end do
800 continue
   ! report error message
   if( err_msg_present(err) ) then
      call err_get_msgs(err, msg_string, platform)
      print "(A)", trim(msg_string)
      print "(A)", "Aborted"
   end if
end program em_test
!####################################################################
```

This is what we see when we run the program from a command line.

```
D:\jls\software\em_exponential>em_test
 Iterations:              278
 Converged: T
 Pi =     0.464573293051795
 Lambda_1 =    0.237951775412863
 Lambda_2 =    0.678085120778212
 Loglik =    -99.3683135052164
 Score:
 -1.934833778779978E-003 -8.597519091892991E-004
 -4.530797434227907E-004
 Hessian:
   -40.3950981345350       87.5777163156822       21.4024109210376
    87.5777163156822     -328.830858103097      -21.9849166839296
    21.4024109210376      -21.9849166839296      -21.7180169385656
```

4.2.8   Test, and Test Again

Our procedure ran, converged, and produced an answer. How can we be sure that this answer is correct? The fact that the first derivatives are close to zero at the solution looks promising. To have greater confidence, however, it would be a good idea to evaluate the loglikelihood function at these estimates independently with a computational package such as S-PLUS, R, or MATLAB. Having a second procedure outside of Fortran for evaluating the loglikelihood, even if it is inefficient, is very useful. With such a procedure, we can perturb the estimated values of π, λ_1, and λ_2 to see if the loglikelihood drops. We can also obtain numerical estimates of derivatives by finite differencing,

$$g'(\theta) \approx \frac{g(\theta + \delta/2) - g(\theta - \delta/2)}{\delta}$$

for a small positive δ. That is, we can perturb each parameter while holding the others fixed to approximate the first and second partial derivatives numerically. This would help us to verify not only that the program is running as intended but that our formulas for the derivatives are correct.

Most Fortran projects will be substantially larger than this one. Our em_exponential_engine module contains one public and one private function and just under 200 lines of source, including comments. In practice, it is unwise to write much more than this without stopping to test what we have written and verify that the results are correct. Novice programmers often make the mistake of writing long programs without stopping to check them along the way. Inevitably, the result is a hodgepodge of errors that are extremely difficult to locate. Any program, aside from a very short one, should be written incrementally and tested repeatedly throughout the coding process. While the computational module is being developed, we will need a method for calling it—usually through a crude console program or a DLL—to systematically test each part before going on to the next step.

4.2.9 Exercises

1. The EM algorithm presented in this section can be modified to perform a constrained maximization of the loglikelihood function. If we fix π equal to some constant value π_0 and do not update its estimate at each cycle, then l will be maximized subject to the constraint $\pi = \pi_0$. The parameters λ_1 and λ_2 may be constrained in a similar fashion. Add optional arguments to `run_em_exponential` that allow a user to constrain any of the parameters to be equal to their starting values.

2. When data are prone to outliers, inferences about location and scale based on a normal population model may be inefficient or misleading, and it makes sense to consider alternatives with heavier tails. Suppose that $y_i = \mu + \sigma \epsilon_i$, $i = 1, \ldots, n$, where the ϵ_i's are drawn from a Student's t distribution with ν degrees of freedom. Little and Rubin (1987) present the following EM algorithm for ML estimation of μ and σ^2 for a fixed ν. Given the current estimates $\mu^{(t)}$ and $\sigma^{2\,(t)}$, calculate weights

$$w_i = \frac{\nu + 1}{\nu + \left(y_i - \mu^{(t)}\right)^2 / \sigma^{2\,(t)}}$$

for $i = 1, \ldots, n$, and update the estimates as

$$\mu^{(t+1)} = \frac{\sum_i w_i y_i}{\sum_i w_i}, \qquad \sigma^{2\,(t+1)} = \frac{1}{n} \sum_i w_i \left(y_i - \mu^{(t+1)}\right)^2.$$

Following the practices used in this section, implement this algorithm in a module procedure. If you are able to compute the log-gamma function (see Exercise 4, Section 4.1.6), evaluate the loglikelihood at each step of EM to make sure that it increases. The density function for y_i is

$$f(y) = \frac{\Gamma\left(\frac{\nu+1}{2}\right)}{\Gamma\left(\frac{\nu}{2}\right)\sqrt{\pi \nu \sigma^2}} \left(1 + \frac{(y-\mu)^2}{\nu \sigma^2}\right)^{-\left(\frac{\nu+1}{2}\right)}.$$

The ML estimate for the degrees of freedom can be found by grid search, examining the loglikelihood achieved for various values of ν.

4.3   Efficient Routines at Lower Levels

4.3.1   What Is a Lower-Level Procedure?

The ML estimation procedure of the previous section is a fairly complete analysis that would be applied only once to a given dataset. One could imagine writing a program that would call `run_em_exponential` repeatedly—for example, when performing a simulation study. Under ordinary circumstances, however, a typical client application is likely to call it just once. Routines of this type are called higher-level procedures. Lower-level procedures, on the other hand, are those that are called from a typical application hundreds, thousands, or millions of times. Examples of lower-level procedures include generating random variates, multiplying or inverting matrices, and computing quantiles or tail areas of common distributions such as normal, t, F, and χ^2.

In describing matrix computations, numerical analysts use more precise notions of level. Level-1 procedures operate on arrays of size n and involve $O(n)$ floating-point operations (e.g., computing an inner product of two vectors), level-2 procedures involve arrays of size mn and do an $O(mn)$ amount of work, and so on. For our purposes, we will not need such a detailed classification; we will simply categorize procedures as lower- or higher-level according to whether they are likely to be called many times or just once. The distinction between these classes is not always clear because a higher-level procedure in one application may serve as a lower-level procedure in another. For example, weighted least-squares regression may be applied once to a dataset, or it may be called repeatedly, as in a procedure for fitting generalized linear models (McCullagh and Nelder, 1989).

Lower-level procedures are the building blocks of a statistical program. A procedure written for one application could easily be used in another. As your programming experience grows, you will want to collect these procedures and arrange them into modules that can be shared among applications. It has often been said that programs tend to follow a 90/10 rule: 90% of the actual processing time is spent executing the instructions from 10% of the source code. The origin of this saying is not clear, although a possible early source is Knuth (1971). Some say that this mythical ratio is closer to 80/20. Either way, the important message is that in computationally intensive programming, the crucial lines of source code are often found at the lower levels. Therefore, increasing efficiency at the lower levels may be the key to improving a program's overall performance. Important strategies for lower-level procedures include reducing the overhead cost associated with each procedure call, taking advantage of special structure, loop reordering, and optimization.

4.3.2 Keeping Overhead Low

Overhead refers to a fixed amount of computation that is performed each time a routine is invoked. When writing a lower-level procedure, it's wise to examine your code for time-consuming tasks that would be redundant if performed each time the procedure is called. If you find any, consider moving them up to procedures at a higher level.

One task that should be avoided in lower-level routines is dynamic allocation of temporary workspaces. If the procedure is called over and over, the repeated allocation and deallocation of memory may seriously degrade performance. Lower-level procedures do sometimes require workspace arrays whose size is not known in advance. If your procedure needs one or more workspaces, it's preferable to pass the workspace as an argument rather than making it a local variable.

For example, consider the problem of approximating a one-dimensional integral by Gauss-Hermite quadrature,

$$\int_{-\infty}^{\infty} e^{-x^2} f(x) dx \approx \sum_{j=1}^{n} w_j f(x_j), \qquad (4.3)$$

where the abscissas $x_1, \ldots, x_n$ and the weights $w_1, \ldots, w_n$ are generated by the theory of orthogonal polynomials. A procedure for computing the x_i's and w_i's for a given n is given by Press et al. (1992, Section 4.5). The accuracy of the approximation may be improved by increasing the number of quadrature points n. An implementation for an arbitrary f might begin like this:

```
integer(our_int) function gauss_hermite( func, n, ans ) &
     result(answer)
   ! n-point Gauss-Hermite quadrature
   implicit none
   ! arguments
   real(our_dble) :: func                  ! the function f
   integer(our_int), intent(in) :: n       ! number of points
   real(our_dble), intent(out) :: ans      ! answer
   ! locals
   real(kind=our_dble), allocatable :: x(:)  ! abscissas
   real(kind=our_dble), allocatable :: w(:)  ! weights
```

The rest of the procedure would allocate x and w to be of size n, compute the abscissas and the weights, compute the right-hand side of (4.3), and deallocate x and w. If this procedure were called repeatedly with the same value for n, the reallocation and recomputation of x and w would be redundant. A better strategy would be to write a separate routine for computing x and w so that these arrays could be created in advance.

Be aware that memory may be allocated within a procedure even if there are no `allocate` statements. In `gauss_hermite`, the two local arrays could have been declared as

```
real(kind=our_dble) :: x(n), w(n)
```

because `n` appears as a dummy argument. In that case, Fortran would have automatically allocated them at the start of the procedure and deallocated them at the end (Section 2.3.5). Arrays of this type, called automatic arrays, should be avoided in lower-level procedures.

There is another situation where an array may be allocated without an explicit request from the programmer: when a dummy argument to a procedure is a fixed-shape or adjustable array but the corresponding actual argument is a pointer. This will be explained shortly, after we discuss issues of array storage and stride.

4.3.3 Taking Advantage of Structure

The essence of computational programming is to convert mathematical formulas into code. Often the formulas provide little or no guidance on how to do this efficiently. In mathematical terms, A^{-1} is well-defined and unique provided that A is nonsingular, yet there are many possible ways to compute A^{-1}. In general, we want to use methods that are tailored to take advantage of any special characteristics of A. It would be wasteful, for example, to apply a general method such as Crout-Doolittle to solve a triangular or banded linear system or to invert a symmetric, positive-definite matrix. Triangular matrices are more effectively inverted by a backsolving operation, banded systems can be solved by a banded version of Gaussian elimination, and the usual way to invert a symmetric, positive-definite matrix is through a Cholesky factorization. Whether you are calling professionally written routines from a package such as LAPACK or writing the routines yourself, be sure to apply methods that take advantage of any special structure that may exist.

Mathematical formulas are designed to be concise, but they often suggest procedures that are grossly inefficient. For example, weighted least-squares regression is conveniently written as $\hat{\beta} = (X^T W X)^{-1} X^T W y$, where W is a matrix with weights $w_1, \ldots, w_n$ on the diagonal and zeroes elsewhere. In practice, it would be extremely wasteful to actually form the $n \times n$ matrix W and compute $X^T W X$ by matrix multiplication. Rather, we would compute the (j, k)th element of $X^T W X$ directly as $\sum_{i=1}^{n} w_i x_{ij} x_{ik}$ for $j \leq k$ and then fill in the lower triangle of this symmetric matrix by copying from the upper triangle. As you convert formulas into code, be aware that concise notation and efficient programming are two entirely different matters and strive to eliminate unnecessary or redundant computation at all levels.

4.3.4   Loop Reordering, Stride and Saxpy

The order in which operations are carried out may have major consequences for efficiency. Consider the problem of computing $AB = C$, where the $(m \times r)$ matrix A and the $(r \times n)$ matrix B have no particular structure. By definition, the (i, j)th element of C is the inner product of the ith row of A and the jth column of B,

$$c_{ij} = \sum_k a_{ik} b_{kj}.$$

The obvious implementation of $AB = C$ is

```
c(:,:) = 0.
do i = 1, m
   do j = 1, n
      do k = 1, r
         c(i,j) = c(i,j) + a(i,k) * b(k,j)
      end do
   end do
end do
```

but this is not the only way. There are six different ways to arrange the computations corresponding to the 3! permutations of the indices i, j, and k. Each involves the same number of floating-point operations, but they do not perform equally well. A better method puts j in the outer loop, k in the middle, and i in the innermost loop:

```
c(:,:) = 0.
do j = 1, n
   do k = 1, r
      do i = 1, m
         c(i,j) = c(i,j) + a(i,k) * b(k,j)
      end do
   end do
end do
```

Why is this second method better? The reasons have to do with the way data are stored and accessed. In Fortran, the elements of an array are stored in a contiguous section of the computer's main memory. The storage order is antilexicographical, in which the subscript corresponding to the first dimension varies fastest, the subscript corresponding to the second dimension varies next fastest, and so on. For example, the elements of A are stored in a columnwise fashion, with a(1,1), a(2,1), ..., a(n,1) followed by a(1,2), a(2,2), ..., a(n,2), etc. Before an array can be operated upon, its contents must be copied from the computer's main memory into the cache, a small bank of high-speed memory locations positioned very close to the processor. If the array is too large to fit in the cache, then it must be moved in one section at a time. This copying of data from one area of memory to another can be more time-consuming than the actual mathematical computations performed by the processor. For efficiency, it helps to arrange the computations to operate on contiguous regions of arrays (e.g., on columns of matrices rather than rows) to reduce the amount of copying to and from the cache.

More generally, the speed of array computations can be greatly affected by stride. Stride refers to the increment or spacing of memory locations that are being operated upon in an iterative procedure. For the array

```
real(kind=our_dble) :: x(p, q, n)
```

an operation on x(i,:,:) has unit stride, an operation on x(:,j,:) has a stride of p, and an operation on x(:,:,k) has a stride of p*q. Computations on arrays, regardless of their size, tend to be most efficient when they are arranged to have unit stride in their innermost loops.

Returning to the problem of computing $AB = C$, consider the innermost loop of the first method:

```
do k = 1, r
   c(i,j) = c(i,j) + a(i,k) * b(k,j)
end do
```

The portions of the arrays A, B and C active in this loop are shown below with the symbol '×',

$$
\begin{bmatrix} \cdot & \cdot & \cdot \\ \times & \times & \times \\ \cdot & \cdot & \cdot \\ \cdot & \cdot & \cdot \\ \cdot & \cdot & \cdot \end{bmatrix}
\begin{bmatrix} \cdot & \times & \cdot & \cdot \\ \cdot & \times & \cdot & \cdot \\ \cdot & \times & \cdot & \cdot \end{bmatrix}
=
\begin{bmatrix} \cdot & \times & \cdot & \cdot \\ \cdot & \cdot & \cdot & \cdot \\ \cdot & \cdot & \cdot & \cdot \\ \cdot & \cdot & \cdot & \cdot \end{bmatrix}.
$$

This operation has a unit stride for B but not for A. In contrast, the innermost loop of the second method,

```
do i = 1, m
   c(i,j) = c(i,j) + a(i,k) * b(k,j)
end do
```

has unit stride for both A and C,

$$
\begin{bmatrix} \cdot & \times & \cdot \\ \cdot & \times & \cdot \\ \cdot & \times & \cdot \\ \cdot & \times & \cdot \\ \cdot & \times & \cdot \end{bmatrix}
\begin{bmatrix} \cdot & \cdot & \cdot & \cdot \\ \cdot & \cdot & \times & \cdot \\ \cdot & \cdot & \cdot & \cdot \end{bmatrix}
=
\begin{bmatrix} \cdot & \cdot & \times & \cdot \\ \cdot & \cdot & \times & \cdot \\ \cdot & \cdot & \times & \cdot \\ \cdot & \cdot & \times & \cdot \\ \cdot & \cdot & \times & \cdot \end{bmatrix}.
$$

This latter operation, which can also be written as

```
c(:,j) = c(:,j) + a(:,k) * b(k,j)
```

is sometimes called a saxpy. Saxpy, an abbreviation for "scalar-alpha x plus y," is an operation of the form $z = \alpha x + y$, where x and y are vectors of equal length and α is a scalar. Many of the matrix procedures commonly used in statistics, including Gaussian elimination and Cholesky factorization, can be arranged to make them rich in saxpys and other unit-stride operations. This is one of the key ideas underlying the high-performance matrix subroutines in BLAS and LAPACK. A guide to matrix computations that exploit the use of saxpys is given by Golub and van Loan (1996).

4.3.5 Optimization, Pipelining and Multiple Processors

When building your application, your compiler may allow you to request varying degrees and types of optimization. Optimization refers to a variety of measures automatically taken to improve overall performance. If your source code contains the statement

```
p = exp(eta) / ( 1. + exp(eta) )
```

most compilers will see the redundant partial expression and implement it as if it had been written like this:

```
tmp = exp(eta)
p = tmp / ( 1. + tmp )
```

Many compilers are capable of analyzing nested loops and identifying where they can be reordered to improve efficiency. Another commonly used trick is loop unrolling, in which iteration cycles of an inner loop are replaced by multiple copies of the code within the body of the loop.

A different set of issues governing efficiency pertains to the way that tasks are scheduled for the computer's processor. Before the processor can perform a task, it must first receive all the details of the instruction, including the basic operation to be performed (e.g., addition), the memory locations of the operands, and where to store the result. The rate at which instructions can be carried out depends not only on the processor's clock speed but on the particulars of its architecture. An old-fashioned processor would wait for an instruction to be fully loaded, execute the instruction, wait for the next one, and so on. Newer, more sophisticated processors are designed for vector computations and pipelining. Pipelining is a technique that increases the overall volume of work performed by reducing the amount of processor idle time, much like an assembly line in a factory. In a pipeline, future instructions are loaded while current instructions are being carried out. Pipelining tends to work best for highly repetitive tasks with unit stride, such as saxpys on long vectors. For shorter vectors, the time required to set up the pipeline may outweigh any resulting savings.

Recent years have also seen developments in parallel computing, in which computational tasks are simultaneously scheduled across multiple processors. The best methods for exploiting parallel architecture typically depend on the details of the system, including the number of processors, the manner in which they communicate with each other and the degree to which they share memory. An extended discussion of how to write Fortran programs to exploit parallel architecture is given by Press et al. (1996).

Most readers will be more interested in writing code that performs well across a variety of systems than optimizing a program for any particular system. Therefore, we suggest that you follow a few simple principles. First, don't automatically assume that the compiler will be smart enough to automatically translate inefficient code into smart code. Take some time to arrange your code to operate on contiguous sections of arrays with unit stride, so that the processor has opportunities to set up efficient pipelines. Be aware of operations on array elements that can be carried out in any order, and implement them with expressions involving whole arrays, array sections, and **where** and **forall** statements. Use intrinsic array functions such as **sum**, **dot_product**, and **matmul** as you wish, but be aware that in some cases these may be slower than writing your own loops (Section 2.2.4). Finally, take advantage of the optimization features of your compiler when your program is ready for distribution and use.

4.3.6   A Simple Example

To illustrate savings that can be achieved from simple rearrangements, consider two different ways of computing $X^T y$, where X is an $n \times p$ matrix and y is an n-vector. The first method operates on rows of X in the innermost loop and has a stride of n:

```
xty(:) = 0.
do i = 1, n
    do j = 1, p
        xty(j) = xty(j) + x(i,j) * y(i)
    end do
end do
```

The second method reverses the loop order and has unit stride:

```
xty(:) = 0.
do j = 1, p
    do i = 1, n
        xty(j) = xty(j) + x(i,j) * y(i)
    end do
end do
```

Defining x, y, and xty as double-precision real arrays, we compiled the two methods with Compaq Visual Fortran Version 6.6 using the default compiler options with no optimization. The programs were run on a 1.59 GHz Pentium M system with 1.00 gigabyte of RAM. With $n = 100,000$ and $p = 200$, the second method ran in 40% less time than the first. When the two methods were recompiled using the highest level of optimization, their performance became identical, presumably because the first pair of loops was automatically reordered.

4.3.7 Hidden Allocation of Temporary Arrays

In old-fashioned FORTRAN, the dimensions of an array dummy argument had to be fixed or had to be passed as arguments themselves. In this procedure,

```
subroutine schmubroutine(a, n, b)
   integer :: n
   real :: a(3), b(n,n)
```

a is called a fixed-shape array, whereas b is said to be an adjustable array. Both types of arrays are still used extensively, and their contents are stored in contiguous memory to help maintain computational efficiency. Similarly, the contents of an allocatable array are stored in contiguous memory. The target of an array pointer, however, is not necessarily contiguous, as you can see from these examples.

```
p1 => a(0:10:2)
p2 => b(i,:)
```

What happens when you pass an array pointer as an actual argument to a procedure whose corresponding dummy argument is a fixed-shape or adjustable array? In most cases, the compiler will create a temporary array in contiguous memory, copy the pointer's target into it, and destroy the temporary array when the procedure has finished. Many Fortran compilers will set up these temporary arrays to maintain contiguity even if they are not really needed (i.e., even if the pointer targets are already contiguous). If the procedure is called in this fashion many times, the repeated allocation and deallocation of temporary arrays can be very inefficient, particularly if the arrays are large.

There are two ways to resolve this problem. First, we can declare the dummy argument to be a pointer rather than a fixed-shape or adjustable array. A disadvantage of that method is that the code within the procedure becomes more difficult to optimize. Moreover, the procedure then becomes less useful to other program units because we can no longer pass non-pointer arrays to it; if a dummy argument is a pointer, the corresponding actual argument must also be a pointer. A better way to resolve this problem is to restructure the calling program so that the actual argument is no longer a pointer but an adjustable or allocatable array. Such revisions are worth considering if the function or subroutine in question is going to be invoked many times; for a higher-level procedure that will be called a few times or just once, it will tend to make little difference.

4.3.8   Exercises

1. One of the most popular techniques for generating normal random variatesnormal random variates is the polar method of Box and Muller (1958). If (v_1, v_2) is uniformly distributed within the circle of unit radius centered at the origin,

$$v_1^2 + v_2^2 \leq 1, \tag{4.4}$$

then

$$z_1 = v_1 \sqrt{\frac{-2\log r^2}{r^2}} \quad \text{and} \quad z_2 = v_2 \sqrt{\frac{-2\log r^2}{r^2}}$$

are independently distributed as $N(0, 1)$, where $r^2 = v_1^2 + v_2^2$. An easy way to simulate a point within the unit circle is to generate v_1 and v_2 uniformly distributed from -1 to 1 and reject them if (4.4) is not satisfied. Implement the polar method in a Fortran procedure that returns a single random variate with each call. Make the procedure as efficient as you can by eliminating unnecessary computation.

2. Section 4.3.6 presented two different ways to compute the matrix-vector product $X^T y$. Implement the first method, which does not have unit stride, and apply it to a large problem using allocatable arrays for x, y, and xty. Measure the speed using the intrinsic function cpu_time (do not include the time required for array allocation), and see how the performance varies as you compile the code with increasing levels of optimization. Then repeat the experiment with x, y, and xty declared as pointers and explain what you find.

3. In the procedure for in-place Cholesky factorization presented in Section 3.6.4, the dummy argument for the matrix had assumed shape. Apply this procedure to a large, positive-definite matrix passed as an allocatable array and time the procedure call using cpu_time. Repeat the experiment, passing the same matrix as an array pointer. Then modify the procedure by making the dummy argument into an adjustable array and repeat the two calls. Discuss what you find. (To create a positive-definite matrix, fill an $n \times p$ matrix X $(n \geq p)$ with random numbers and compute $X^T X$.)

4.4   More Computational Procedure Examples

4.4.1   An Improved Cholesky Factorization

We now demonstrate some of the principles described in this chapter by writing a few lower-level matrix routines and a higher-level procedure that calls them. These procedures are not of professional quality; when very high speed and extreme precision are required, there is no substitute for code written by experienced numerical analysts. Our goal here is to create routines that are accurate and efficient enough for most applications while keeping the source code clean and simple.

In Section 3.6.4, we presented a function that overwrites the lower triangle of a symmetric, positive-definite matrix with its Cholesky factor. Examining this procedure, we find that its innermost loop,

```
do k = 1, j-1
   sum = sum + a(j,k)*a(i,k)
end do
```

runs through rows of the matrix and therefore does not have unit stride. An alternative algorithm described by Golub and van Loan (1996, Algorithm 4.2.1) arranges the computations to be rich in saxpy operations. An implementation of that algorithm is shown below.

```
_____ matrix.f90 _____

!###################################################################
integer(kind=our_int) function cholesky_saxpy(a, err) result(answer)
   !### Overwrites lower triangle of a symmetric, pos.-def.
   !### matrix a with its Cholesky factor.
   implicit none
   ! declare arguments
   real(kind=our_dble), intent(inout) :: a(:,:)
   type(error_type), intent(inout) :: err
   ! declare local variables and parameters
   character(len=*), parameter :: subname = "cholesky_saxpy"
   integer(kind=our_int) :: p, j, k
   real(kind=our_dble) :: den
   ! begin
   answer = RETURN_FAIL
   p = size(a,1)
   if( p /= size(a,2) ) goto 700
   do j = 1, p
      do k = 1, j-1
         a(j:p,j) = a(j:p,j) - a(j:p,k) * a(j,k)
      end do
      if( a(j,j) <= 0.D0 ) goto 710
      den = sqrt( a(j,j) )
      a(j:p,j) = a(j:p,j) / den
   end do
   ! normal exit
   answer = RETURN_SUCCESS
   return
   ! error traps
700 call err_handle(err, 300, &
        called_from = subname//" in MOD "//modname)
   return
710 call err_handle(err, 101, &
        called_from = subname//" in MOD "//modname)
   return
```

```
end function cholesky_saxpy
!####################################################################
```

 Style tip

Use assumed-shape rather than adjustable arrays as dummy arguments, and don't give them the **pointer** attribute unless it is necessary (e.g., if they need to be resized within the procedure). If the procedure will be public, add safeguards to ensure that the actual arguments have the correct dimensions.

Note that `cholesky_saxpy` uses the programming constants defined in Section 4.2.2. We have placed it in a module called `matrix_methods` along with the two functions shown next.

4.4.2 Inverting a Symmetric Positive-Definite Matrix

When solving a linear system $Ax = b$, it is best to compute $x = A^{-1}b$ without explicitly finding A^{-1}. In statistical computation, however, the inverse of a matrix is often needed, especially if the matrix is symmetric and positive definite (SPD). The standard way to invert SPD matrices is by the Cholesky factorization: take $A = CC^T$, then $A^{-1} = B^TB$, where $B = C^{-1}$. A nice by-product of this method is that the determinant of A is available from the diagonal elements of the Cholesky factor, $|A| = \prod_i c_{ii}^2$.

The inverse of a lower-triangular matrix, which is also lower triangular, may be computed by a variant of forward substitution. A procedure for overwriting a lower triangle with its inverse is shown below. In the innermost loop, this procedure accesses the elements of the matrix by rows and by columns. Without overwriting, a version of forward substitution is available based entirely on columns (Golub and van Loan, 1996).

```
_____ matrix.f90 _____
!####################################################################
integer(kind=our_int) function invert_lower(a, err) result(answer)
   !### Overwrites a lower-triangular matrix a with its inverse
   !### by forward substitution. The upper triangle is untouched.
   implicit none
   ! declare arguments
   real(kind=our_dble), intent(inout) :: a(:,:)
   type(error_type), intent(inout) :: err
   ! declare local variables and parameters
   character(len=*), parameter :: subname = "invert_lower"
   integer(kind=our_int) :: p, i, j, k
   real(kind=our_dble) :: sum
   ! begin
   answer = RETURN_FAIL
   p = size(a,1)
```

```
      if( p /= size(a,2) ) goto 700
      if( a(1,1) == 0.D0 ) goto 710
      a(1,1) = 1.D0 / a(1,1)
      do i = 2, p
         if( a(i,i) == 0.D0 ) goto 710
         a(i,i) = 1.D0 / a(i,i)
         do j = 1, i-1
            sum = 0.D0
            do k = j, i-1
               sum = sum + a(k,j) * a(i,k)
            end do
            a(i,j) = - sum * a(i,i)
         end do
      end do
      ! normal exit
      answer = RETURN_SUCCESS
      return
      ! error traps
700   call err_handle(err, 300, &
           called_from = subname//" in MOD "//modname)
      return
710   call err_handle(err, 100, &
           called_from = subname//" in MOD "//modname)
      return
   end function invert_lower
   !###################################################################
```

 Style tip ──

If you do not want to overwrite the matrix supplied, do not use this function. Rather, overload it with a companion function that accepts two array arguments, one that is intent(in) and another that is intent(out). Having both procedures available helps to avoid unnecessary copying of data.

───

The final step of inversion is to premultiply a lower triangle by its transpose. Here is a multiplication procedure that avoids unnecessary computation by skipping elements known to be zero and by capitalizing on the symmetry of the result. The input array is accessed entirely by column.

──────────────── matrix.f90 ────────────────

```
   !###################################################################
   integer(kind=our_int) function premult_lower_by_transpose(a, b, &
        err) result(answer)
      !### Premultiplies a lower-triangular matrix a by its upper-
      !### triangular transpose to produce a symmetric matrix b.
      implicit none
      ! declare arguments
      real(kind=our_dble), intent(in) :: a(:,:)
      real(kind=our_dble), intent(out) :: b(:,:)
      type(error_type), intent(inout) :: err
```

```
    ! declare local variables and parameters
    character(len=*), parameter :: subname = &
        "premult_lower_by_transpose"
    integer(kind=our_int) :: p, i, j, k
    ! begin
    answer = RETURN_FAIL
    p = size(a,1)
    if( p /= size(a,2) ) goto 700
    if( ( p /= size(b,1) ) .or. ( p /= size(b,2) ) ) goto 710
    do i = 1, p
       do j = 1, i
          b(i,j) = 0.D0
          do k = max(i,j), p    ! skip zero elements
             b(i,j) = b(i,j) + a(k,i) * a(k,j)
          end do
          b(j,i) = b(i,j)       ! copy upper triangle from lower
       end do
    end do
    ! normal exit
    answer = RETURN_SUCCESS
    return
    ! error traps
700 call err_handle(err, 300, &
        called_from = subname//" in MOD "//modname)
    return
710 call err_handle(err, 301, &
        called_from = subname//" in MOD "//modname)
    return
  end function premult_lower_by_transpose
  !#################################################################
```

4.4.3   Weighted Least Squares

To illustrate the use of these lower-level procedures, we now develop a computational routine for weighted least-squares (WLS) regression of a vector $y = (y_1, \ldots, y_n)^T$ on a full-rank $n \times p$ matrix X. The solution to the ordinary least-squares (OLS) problem, $\hat{\beta} = (X^T X)^{-1} X^T y$, can be computed in a variety of ways. A procedure recommended by numerical analysts for its stability and efficiency involves transforming the columns of X by Householder rotations, which eliminates the need to compute and invert $X^T X$; a good discussion of that method is provided by Thisted (1988). When weights $w_1, \ldots, w_n$ are present, the solution becomes

$$\hat{\beta} = (X^T W X)^{-1} X^T W y,$$

where $W = \text{Diag}(w_1, \ldots, w_n)$; this is equivalent to performing OLS on the transformed variables $y^* = W^{1/2} y$ and $X^* = W^{1/2} X$, where $W^{1/2} = \text{Diag}(\sqrt{w_1}, \ldots, \sqrt{w_n})^T$. Except in very extreme cases—e.g., when the regressors are almost collinear or when the weights vary by many orders of magnitude—a simpler method using the Cholesky-based inversion of $X^T W X$ works well enough. Explicitly computing $(X^T W X)^{-1}$ may also be necessary for statistical inferences because the unbiased estimate of $V(\hat{\beta})$ is $\hat{\sigma}^2 (X^T W X)^{-1}$, with

$$\hat{\sigma}^2 = \frac{1}{n-p} (y - X\hat{\beta})^T W (y - X\hat{\beta}).$$

Here's a procedure for WLS that returns $\hat{\beta}$, $(X^T W X)^{-1}$, and $\hat{\sigma}^2$.

─────────── wls.f90 ───────────

```
!###################################################################
integer(kind=our_int) function fit_wls(x, y, w, beta, &
     cov_unscaled, scale, err) result(answer)
   ! Regresses y on x, using weights in w.
   !    beta = estimated coefficients
   !    cov_unscaled = inverse of (X^T W X) matrix
   !    scale = unbiased estimate of scale parameter sigma^2
   implicit none
   ! declare arguments
   real(kind=our_dble), intent(in) :: x(:,:), y(:), w(:)
   real(kind=our_dble), intent(out) :: beta(:), &
        cov_unscaled(:,:), scale
   type(error_type), intent(inout) :: err
   ! declare local variables and parameters
   character(len=*), parameter :: subname = "fit_wls"
   real(kind=our_dble), allocatable :: xtwx(:,:), xtwy(:)
   integer(kind=our_int) :: n, p, i, j, k, status
   ! check arguments
   answer = RETURN_FAIL
   n = size(x,1)
   p = size(x,2)
   if( ( size(y) /= n ) .or. ( size(beta) /= p ) .or. &
       ( size(w) /= n) .or. ( size(cov_unscaled,1) /= p ) .or. &
       ( size(cov_unscaled,2) /= p ) ) goto 700
   ! form xtwy and lower triangle of xtwx
   allocate( xtwx(p,p), xtwy(p), stat=status)
   if( status /= 0 ) goto 710
   do j = 1, p
      xtwy(j) = sum( x(:,j) * w(:) * y(:) )
      do k = 1, j
         xtwx(j,k) = sum( x(:,j) * w(:) * x(:,k) )
      end do
   end do
   ! compute cov_unscaled and beta
   if( cholesky_saxpy( xtwx, err ) == RETURN_FAIL ) goto 800
   if( invert_lower( xtwx, err ) == RETURN_FAIL ) goto 800
   if( premult_lower_by_transpose( xtwx, cov_unscaled, err ) &
        == RETURN_FAIL ) goto 800
   beta = matmul( cov_unscaled, xtwy )
   ! compute scale
   scale = 0.D0
   do i = 1, n
```

```
            scale = scale + w(i) * ( y(i) - sum( x(i,:) * beta ) )**2
         end do
         scale = scale / real( n - p, kind=our_dble )
         ! deallocate workspaces
         deallocate(xtwx, xtwy, stat=status)
         if( status /= 0 ) goto 720
         ! normal exit
         answer = RETURN_SUCCESS
         return
         ! error traps
700      call err_handle(err, 301, &
              called_from = subname//" in MOD "//modname)
         return
710      call err_handle(err, 200, &
              called_from = subname//" in MOD "//modname)
         return
720      call err_handle(err, 201, &
              called_from = subname//" in MOD "//modname)
         return
800      call err_handle(err, 1000, &
              called_from = subname//" in MOD "//modname)
         deallocate(xtwx, xtwy)
         return
      end function fit_wls
      !##############################################################
```

Notice the use of local allocatable arrays for holding $X^T W X$ and $X^T W y$. If this procedure is called many times, the repeated allocation might become costly, particularly if the number of regressors p is large. In that case, we might want to pass the two workspaces as arguments and require the calling program to allocate them.

4.4.4   Computational Routines in Object-Oriented Programming

Our WLS procedure, with the exception of its error-handling features, does not rely upon any special data structures; all of the inputs and outputs are arrays and scalar variables. Principles of object-oriented programming, however, suggest that we may want to bundle x, y, and w together as a derived type, allowing us to pass all the input data through a single argument. Similarly, we could collect beta, cov_unscaled, and scale into another derived type, so that all the results are returned through a single argument. Why have we chosen not to apply more derived types to this problem and to the other computational routines in this chapter?

One reason why we avoided derived types as arguments in this chapter is to enhance efficiency. Derived types usually contain pointers, and unnecessary use of pointers can make computational routines slow and difficult to optimize.

Another reason we avoided derived types is that we like to keep all arguments to computational routines as simple and general as possible, so that they can be more easily shared among applications. Suppose that we had begun our WLS procedure like this:

```
integer(kind=our_int) function fit_wls( wls_data, &
    wls_results, err) result(answer)
    implicit none
    ! declare arguments
    type(wls_data_type), intent(in) :: wls_data
    type(wls_results_type), intent(out) :: wls_results
    type(error_type), intent(inout) :: err
```

Where would the definitions of `wls_data_type` and `wls_results_type` reside? They would have to be placed either in this same module or in another module used by this one. Either way, a program unit that invoked `fit_wls` would have to supply and retrieve data through these types. That could become very inconvenient if, at some future time, we tried to use this function as a building block for a more complicated procedure (e.g., fitting a logistic regression model by iteratively reweighted least squares).

Designing specialized data structures is an excellent idea, but we prefer to keep them hidden from our computational routines. Modules that define object classes and methods should use the computational routines rather than the other way around. Suppose that we were designing an object class for weighted linear regression analysis. We would begin by creating a module that contains definitions for `wls_data_type` and `wls_results_type`. We would write public methods for putting data into these types and for getting results from them, as described in Chapter 3. Then we would write a public method, say `run_wls_regression`, that would perform the computations by invoking `fit_wls`. The `run_wls_regression` procedure would be essentially a wrapper for `fit_wls`, perhaps with some extra safeguards to prevent misuse (e.g., trying to fit a model before the data are supplied). At first, writing this additional code might seem like a waste of time. In the long run, however, you will find that it saves time by promoting reuse of the computational routines.

 Style tip————————————————————————————————

Avoid using specialized data types as arguments to your computational routines.

4.5   Additional Exercises

1. The `cholesky_saxpy` and `invert_lower` procedures developed in Section 4.4 overwrite the input matrix with the result. Overload these with companion functions that do not alter the input but write the result to another matrix.

2. Modify `fit_wls` to make the weight argument optional. If no weights are supplied, it should assume $w_i = 1$ for all i and perform ordinary least squares.

3. Modify `fit_wls` to allow a user to optionally supply workspaces for $X^T W X$ and $X^T W y$ so that these arrays do not need to be allocated within the procedure. Do this without using array pointers. (Hint: Try overloading.)

4. The sweep operator is a useful device for stepwise regression and linear model computations (Little and Rubin, 1987; Thisted, 1988). SWP[k] operates on a symmetric $p \times p$ matrix G by replacing it with another $p \times p$ symmetric matrix H,

$$H = \mathrm{SWP}[k]\, G,$$

whose elements are

$$
\begin{aligned}
h_{kk} &= -1/g_{kk}, \\
h_{jk} &= h_{kj} = g_{jk}/g_{kk} \text{ for } j \neq k, \\
h_{jl} &= h_{lj} = g_{jl} - g_{jk}g_{kl}/g_{kk} \text{ for } j \neq k \text{ and } l \neq k.
\end{aligned}
$$

   a. Write a Fortran procedure that overwrites the lower triangle of G with SWP[k] G.

   b. Successively sweeping the matrix G on positions $1, 2, \ldots, p$ replaces it with $-G^{-1}$. The sweeps may be carried out in any order, and G^{-1} exists if and only if none of the sweeps involve attempted division by zero. Using sweep, write a procedure that overwrites the lower triangle of a symmetric matrix with its inverse.

5. Suppose that $y_1, y_2, \ldots, y_n$ are a random sample from a p-variate normal distribution $N(\mu, \Sigma)$. Collect the entire sample into an $n \times p$ data matrix Y whose rows are y_i^T, $i = 1, \ldots, n$.

   a. Write a procedure for calculating the maximum-likelihood (ML) estimates

$$\hat{\mu} = \bar{y} = n^{-1} \sum_{i=1}^{n} y_i$$

and

$$\hat{\Sigma} = n^{-1}Y^TY - \bar{y}\bar{y}^T$$
$$= n^{-1}\sum_{i=1}^{n}(y_i - \bar{y})(y_i - \bar{y})^T.$$

b. Write another Fortran procedure that evaluates the loglikelihood function

$$l(\mu, \Sigma) = -\frac{np}{2}\log(2\pi) - \frac{n}{2}\log|\Sigma|$$
$$-\frac{1}{2}\sum_{i=1}^{n}(y_i - \mu)^T\Sigma^{-1}(y_i - \mu)$$

at any μ and Σ. For your procedure, allow the user to supply either the raw data Y or the sufficient statistics $T_1 = \sum_i y_i$ and $T_2 = Y^TY$. An identity that you may find useful is

$$\sum_{i=1}^{n}(y_i - \mu)^T\Sigma^{-1}(y_i - \mu) = \text{tr}\,\Sigma^{-1}\left[\sum_{i=1}^{n}(y_i - \mu)(y_i - \mu)^T\right].$$

c. Using a dataset of your choosing, verify that $\hat{\mu}$ and $\hat{\Sigma}$ are indeed the ML estimates by perturbing the elements of these parameters one at a time to see if the loglikelihood drops.

6. Consider the usual regression problem $y_i = x_i^T\beta + \epsilon_i$, $i = 1, \ldots, n$, where y_i is a real-valued response, x_i is a $p \times 1$ vector of predictors, and $\epsilon_1, \ldots, \epsilon_n$ are independent and identically distributed with mean zero. The OLS estimate for β is optimal if the errors are normally distributed but can be inefficient if the data are prone to outliers. In particular, if

$$\epsilon_i \sim \sigma t_\nu,$$

where t_ν denotes a Student's t random variable with ν degrees of freedom, a better estimate of β can be found by an extension of the EM algorithm outlined in Exercise 2 of Section 4.2.9. Given the current estimates $\beta^{(t)}$ and $\sigma^{2\,(t)}$, we compute the weights

$$w_i = \frac{\nu + 1}{\nu + (y_i - x_i^T\beta^{(t)})^2/\sigma^{2\,(t)}}$$

for $i = 1, \ldots, n$, and then update our estimates of β and σ^2 by weighted least squares. (Because this is an ML procedure, the estimate of σ^2 should use a denominator of n rather than $n - p$.) Implement this EM algorithm as a Fortran procedure that calls `fit_wls`. Use OLS estimates as starting values.

5

Developing a Console Application

Accurate and efficient numerical routines are the heart of a quality statistical program. Many lines of code are needed, however, to turn a computational procedure into a useful tool for data analysis. The tasks performed by this additional code—receiving instructions from the user, opening and reading data files, printing results, and so on—are often considered mundane. But if this infrastructure is shoddy, the application will not be very useful, no matter how good the numerical procedures are.

This chapter is a step-by-step guide to writing statistical programs in Fortran that are invoked from a command line. Command-line programs, also known as console applications, may seem crude or outdated in today's computing environments. Nevertheless, they have many important advantages. They are highly portable; with essentially no changes to the source code, they can be compiled with any implementation of Fortran and produce equivalent results on Windows, Unix, Linux, or Macintosh systems. Once compiled, the executable file can be used by anyone without the purchase of additional software.

Using the example of logistic regression, we demonstrate how to write a console program in a modular, pseudo object-oriented fashion, paying careful attention to the overall architecture and interface. Statistical programmers who follow this template will find that their code is easy to debug, maintain, and extend, and that many of the software components can be reused in other applications with little or no modification. Moreover, a console program written in this style is more easily called from another project, repackaged as a COM server, or connected to a graphical user interface using the methods described in later chapters.

5.1   A Program for Logistic Regression

5.1.1   The logistic regression model

Logistic regression is the most popular statistical model for binary data. The response variable for the ith observational unit, y_i, is a count of how many "successes" occurred in n_i "trials." We assume that y_i follows a binomial distribution, $y_i \sim \mathrm{Bin}(n_i, \pi_i)$, $i = 1, \ldots, N$, where n_i is known and the success rate π_i depends on known covariates $x_i = (x_{i1}, \ldots, x_{ip})^T$. The assumed relationship between the success rate and the covariates is

$$\mathrm{logit}(\pi_i) = \log\left(\frac{\pi_i}{1 - \pi_i}\right) = x_i^T \beta, \tag{5.1}$$

where $\beta = (\beta_1, \ldots, \beta_p)$ is a vector of unknown coefficients to be estimated. In most cases, we define $x_{i1} \equiv 1$ so that the first element of β becomes an intercept. The linear predictor $x_i^T \beta$ can be interpreted as the log-odds of success, and $\exp(\beta_j)$ is the amount by which the odds are multiplied when x_{ij} increases by one unit.

Maximum-likelihood estimates of the unknown parameters β are typically computed by a Newton-Raphson or Fisher scoring procedure that can be viewed as iteratively reweighted least-squares regression. For more discussions of logistic regression and related models for dichotomous outcomes, see McCullagh and Nelder (1989), Hosmer and Lemeshow (2000), or Agresti (2002).

5.1.2   Motivation for the ELOGIT Console Program

Although software for fitting the logistic model is widely available, we have developed a new console application for logistic regression called ELOGIT to illustrate principles of statistical software design. Logistic regression provides an excellent example because it is familiar to most readers, the computations are simple but not trivial, and many of the subtasks required— reading in a rectangular dataset, specifying the model to be fit, obtaining starting values for the parameters, iterating to a final solution, and summarizing the results—are ubiquitous in the statistical world. The open-source ELOGIT program provides a template for more complicated programming tasks, and many of its components can be reused with little or no modification in other programs. Readers are encouraged to download the source-code components from our Web site to use, adapt, and extend them for their own purposes.

5.1.3   Dependency Map for Source Components

The source for the main ELOGIT program has only a few dozen lines; all other code resides in modules. A diagram of these modules showing

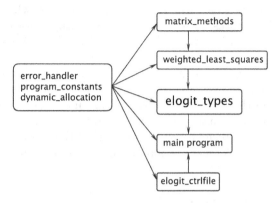

FIGURE 5.1. Dependencies among major components in the ELOGIT console program.

the dependencies among them is shown in Figure 5.1. In this diagram, an arrow from one unit to another, as in $A \to B$, indicates that unit A provides services to B (i.e., is used by B). A double-headed arrow as in $A \leftrightarrow B$ is not possible because two Fortran modules cannot use each other.

The left-hand side of Figure 5.1 shows a group of three modules that are used pervasively: error_handler (which was presented in Section 3.6), program_constants, and dynamic_allocation. The largest and most important module is elogit_types; it defines an object class for storing the data, model specification, parameters, and results, and it contains all of the puts, gets, and other methods associated with this class. Computational services are furnished by weighted_least_squares and matrix_methods, which were developed in Section 4.4. The module elogit_ctrlfile handles the reading of user-supplied instructions from a control file.

Many other designs could work well for this particular application. In our experience, however, we have found that an arrangement such as this produces a program that is easy to debug and extend, it allows components to be readily reused and adapted for other programs, and it greatly simplifies the process of converting the console application to a COM server, as we will describe in Chapter 7. When converting to a COM server, only one small module (elogit_ctrlfile) is discarded, and the functions of the main program are taken over by an external client written in Excel, SAS, S-PLUS, or some other system.

5.1.4 Developing Program Units Incrementally

Dependencies among program units require them to be compiled in a certain order. If module A provides services to B, it must be compiled before B. This does not mean, however, that module A must be completely finished before B is written. In practice, we do not know whether a public module procedure works until another module actually uses it. If module A is used by B, it helps to develop both modules simultaneously along with the main program, introducing new features and testing them incrementally at each step. The dependency map in Figure 5.1 says little about where to begin and how to proceed. Therefore, we describe the development of ELOGIT from the start, discussing options and providing the rationale for each major decision along the way. In addition to the full source code, we present on our Web site a few "snapshots" of the ELOGIT program in various stages of development as major milestones are achieved.

5.2   Where to Begin

5.2.1   Before You Start

It is unwise to start coding before you articulate what the program will do. ELOGIT, a simple console program for logistic regression, will:

- read in a data file supplied by the user;

- determine the model to be fit (i.e., which variable is the response, which are the predictors, whether an intercept is to be included);

- carry out the procedure for maximum-likelihood estimation; and

- write a nicely formatted summary of results—including the estimated coefficients, standard errors, loglikelihood, and fit statistics—to a file.

This description omits many important details, but it's good enough to get us started. We will build up the program incrementally, developing and testing the functionality of each task in the order listed above.

Before starting a larger, more elaborate project—particularly one that will be developed by multiple programmers—it makes sense to prepare an official statement of work (SOW) document that lists the program requirements. The SOW becomes an invaluable reference for the project team, helping to ensure that components developed by different persons will ultimately work together.

One who is going to do a lot of programming, either individually or as part of a team, should also seriously consider adopting a formal system for source-code management. Tools for code management—also known as version-control or revision-control systems—provide a degree of safety by keeping the SOW, code files, documentation, and other materials in a central repository where they can be regularly backed up. An Internet-based management system gives multiple programmers immediate access to the latest version of any file while ensuring that no two persons will be modifying a file at the same time. A good system will also allow programmers

to step backward, obtaining copies of project files as they appeared at any
date and time in the past.

Even if you intend to do all the programming by yourself, formal procedures for code management are still important. Most applications that you
write will have some components that are specific to that application and
others that are shared among multiple applications. If multiple versions
of a shared source-code file exist on your computer, it's not easy to keep
them all synchronized and up-to-date. Many free and commercial programs
for source-code management are available to help you keep track of your
various projects and files; acquiring one of these tools and learning how to
use it is well worth the effort.

5.2.2 Program Constants

As usual, this software project begins with a module of named constants.
These include kind parameters for integer and real variables and values
for RETURN_SUCCESS and RETURN_FAIL. To these we have added Fortran
I/O unit numbers for reading and writing to various types of external files.
Defining I/O unit numbers as named constants rather than literal integers
helps to avoid conflicts that may arise if multiple files are open simultaneously. We have also included character-string constants containing the
program's name, authors, version number, build date, and so on. These
strings can be used within the program for generating messages and stamp-
ing output files. At this point, we don't need to anticipate all the constants
that will be needed for the final program because more can be added at
any time. We created this constants module by making a few changes to
one from an earlier program.

```
───────────────────── constants.f90 ─────────────────────
!#####################################################################
module program_constants
    ! Programming constants used throughout the ELOGIT program.
    ! Unlike most modules, everything here is public.
    implicit none
    public
    ! Define compiler-specific KIND numbers for integers,
    ! single and double-precision reals to help ensure consistency of
    ! performance across platforms:
    integer, parameter :: our_int = selected_int_kind(9), &
        our_sgle = selected_real_kind(6,37), &
        our_dble = selected_real_kind(15,307)
    ! Define UNIT numbers for Fortran I/O:
    integer, parameter :: ctrl_file_handle = 11
    ! Define maximum lengths for various types of character strings:
    integer, parameter :: file_name_length = 256
    ! Define the maximum line widths for various types of files:
    integer, parameter :: ctrl_line_width = 80
    ! Common integer values returned by all functions to indicate
    ! success or failure:
```

```
    integer(kind=our_int), parameter :: RETURN_SUCCESS = 0, &
        RETURN_FAIL = -1
    ! Character strings describing this program:
    character(len=*), parameter :: &
        program_name = "ELOGIT", &
        program_description = &
        "A simple program for logistic regression analysis", &
        program_version = "Version 1.0", &
        program_version_and_date = "Version 1.0 - June, 2004", &
        program_author = "Written by J.L. Schafer", &
        program_institution_1 = &
        "Department of Statistics and The Methodology Center", &
        program_institution_2 = "The Pennsylvania State University"
end module program_constants
!##################################################################
```

5.2.3   The Control File Interface

Our next task is to devise a mechanism by which the user communicates run-time options to the program. With simple programs (e.g., the uniform generators in Chapter 2), these options can be input by the user at the command line as responses to a running dialog. As the number of options grows, however, it becomes easier for the user to enter all the responses into a control file and then run the program by typing a single command.

A control file interface does not have to be crude; it can be made surprisingly convenient and easy to use by including a few simple features. One common problem with control files is that users may have difficulty keeping track of line numbers, forgetting which options need to go on which lines. To ameliorate this problem, we will divide the ELOGIT control file into short sections, allowing each section to begin with an arbitrary number of comment lines. We will also maintain a line counter; if the file contains a mistake, we can then provide an error message, reporting to the user the exact line in the control file where the problem occurred.

We now begin to write the elogit_ctrlfile module. This module has two main parts: the definition of a derived type to organize all the responses from the control file into a single object, and a function that opens the control file and reads the responses. What should go into the derived type? Recalling our mission statement, the first major task for ELOGIT is to read a dataset from a file. We haven't yet decided how the data file will look, but we can make a few decisions now. First, we anticipate that the data will look like a rectangular matrix, with rows corresponding to subjects or cases and columns corresponding to variables. Next, we will allow (but not require) the user to supply names for the variables via a separate file distinct from the data file. Finally, we will also allow (but again not require) the first field in each line of the data file to serve as a case identifier. We can now begin to create the derived type to store control file responses.

—————————————— elogit_ctrlfile.f90 ——————————————

```
!####################################################################
type :: elogit_ctrlfile_type
   ! unlike other types, the contents of this one are public
   sequence
   ! data input section
   integer(kind=our_int) :: ncase=0, nvar=0
   logical :: case_id_present=.false., names_file_present=.false.
   character(len=file_name_length) :: data_file_name="", &
        names_file_name=""
   !
   ! We'll add more components to this type later
   !
end type elogit_ctrlfile_type
!####################################################################
```

Whenever we create a new type, we write a procedure for returning all
components to their initialized states. If any of the components are pointers,
then this procedure should deallocate them.

—————————————— elogit_ctrlfile.f90 ——————————————

```
!####################################################################
integer(our_int) function nullify_elogit_ctrlfile(ctrlfile, err) &
     result(answer)
   ! Returns an elogit_ctrlfile_type to its initialized (null)
   ! state
   implicit none
   ! declare arguments
   type(elogit_ctrlfile_type), intent(inout) :: ctrlfile
   type(error_type), intent(inout) :: err
   ! begin
   ctrlfile%ncase = 0
   ctrlfile%nvar = 0
   ctrlfile%case_id_present = .false.
   ctrlfile%names_file_present = .false.
   ctrlfile%data_file_name = ""
   ctrlfile%names_file_name = ""
   !
   ! Add more code later, as more components are added to the type
   !
   ! normal exit
   answer = RETURN_SUCCESS
end function nullify_elogit_ctrlfile
!####################################################################
```

 Style tip ——

For each derived type, write a procedure that reinitializes its components.

———

Let's also begin a document that specifies the format of the control file.
It's easiest to create this document now while we develop the code.

```
━━━━━━━━━━━━━━━━━━━━ controlfilespec.txt ━━━━━━━━━━━━━━━━━━━
FORMAT OF THE ELOGIT CONTROL FILE
---------------------------------
COMMENT LINES: An unlimited number of comment lines is allowed at the
beginning of the control file, and between each of the sections
below. Each comment line must begin with (i.e. the first nonblank
character on the line must be) an asterisk "*". These comment lines
are ignored.
*******************************************************************
* Data input section
* Optional comment lines, followed by:
LINE 1:       ncase, nvar, case_id_present
        ncase = number of cases or rows in the dataset
        nvar = number of variables (excluding case id, if present)
        case_id_present = T if the first field on each line of the
             data file is to be interpreted as a case identifier
             string rather than a variable
LINE 2:       data_file_name
        character string giving the name of ASCII data file
LINE 3: names_file_name
        character string giving the name of the optional names file
        If no names file is provided, leave this line blank.
*******************************************************************

* We'll add more information to this file later
```

 Style tip ───

Create important documentation files while you are writing the code.

Now let's begin to write a function that reads the control file. Errors in
opening files may cause a program to crash, but these may be prevented
by specifiers in the **open** statement. If a file is to be opened for reading,
status='OLD' throws an error if the file is not found or cannot be opened
for any reason (e.g., if it is currently locked by another application). If an
error occurs, control can be diverted to a labeled statement through the **err**
specifier. The **read** statement also has useful specifiers; **eof**, **eor**, and **err**
will transfer control to a labeled statement if an end-of-file, end-of-record,
or error condition is encountered. The function below uses another function,
skip_comment_lines, which is not shown. To see how that function works,
open the file **elogit_ctrlfile.f90** and examine it for yourself.

```
━━━━━━━━━━━━━━━━━━━━ elogit_ctrlfile.f90 ━━━━━━━━━━━━━━━━━━━
!################################################################
integer(kind=our_int) function read_elogit_ctrlfile( &
     control_file_name, ctrlfile, err) result(answer)
   ! Reads information from an ELOGIT control file and stores
   ! it in an elogit_ctrlfile_type. If the read operation fails for
   ! any reason, the elogit_ctrlfile_type is nullified and the
```

```
! returned value is RETURN_FAIL.
implicit none
! declare arguments
character(len=*), intent(in) :: control_file_name
type(elogit_ctrlfile_type), intent(out) :: ctrlfile
type(error_type), intent(inout) :: err
! declare local variables and parameters
integer(kind=our_int) :: current_line, ijunk
character(len=ctrl_line_width) :: line
character(len=*), parameter :: subname = "read_elogit_ctrlfile"
! open the control file
answer = RETURN_FAIL
if(control_file_name == "") goto 700
open(unit=ctrl_file_handle, file=control_file_name, &
     status="old", err=800)
current_line = 0
!#############################
! Dataset input section
if( skip_comment_lines(ctrl_file_handle, current_line) &
     == RETURN_FAIL ) goto 900
! read ncase, nvar, case_id_present
current_line = current_line + 1
read(unit=ctrl_file_handle, fmt="(A)", err=900, end=900) line
read(line, *, err=900, end=900) &
     ctrlfile%ncase, ctrlfile%nvar, ctrlfile%case_id_present
! read data_file_name and left-justify
current_line = current_line+1
read(unit=ctrl_file_handle, fmt="(A)", err=900, end=900) &
     ctrlfile%data_file_name
ctrlfile%data_file_name = adjustl(ctrlfile%data_file_name)
! read names_file_name and set names_file_present
current_line = current_line+1
read(unit=ctrl_file_handle, fmt="(A)", err=900, end=900) &
     ctrlfile%names_file_name
if( ctrlfile%names_file_name == "" ) then
   ctrlfile%names_file_present = .false.
else
   ctrlfile%names_file_present = .true.
   ctrlfile%names_file_name = adjustl(ctrlfile%names_file_name)
end if
!#############################

! We'll add more code here later

! normal exit
close(unit=ctrl_file_handle)
answer = RETURN_SUCCESS
return
! error traps
700  call err_handle(err, 1000, &
        called_from = subname//" in MOD "//modname, &
        custom_1 = "No control file name specified.")
     goto 999
```

```
800    call err_handle(err, 1, &
             called_from = subname//" in MOD "//modname, &
             file_name = control_file_name)
       goto 999
900    call err_handle(err, 3, &
             called_from = subname//" in MOD "//modname, &
             file_name = control_file_name, line_no = current_line)
       goto 999
       ! final cleanup in the event of an error
999    continue
       close(unit=ctrl_file_handle)
       ijunk = nullify_elogit_ctrlfile(ctrlfile, err)
     end function read_elogit_ctrlfile
     !###############################################################
```

 Style tip

Use named constants rather than literal integers for file unit numbers. Use specifiers such as **err** to manage errors in **open** and **read** statements.

Errors during execution of a **read** statement can also be handled through the **iostat** specifier, which sets an integer variable to zero if the operation was successful or to a positive value depending on the type of error that occurred. We prefer not to use **iostat**, however, because the error codes are not standardized and vary from one compiler to another.

Notice that the line containing **ncase**, **nvar**, and **case_id_present** is not read directly from the control file into those three variables. Rather, the entire line is first read into a character-string buffer, and the data items are then read from that buffer. Why have we done this? Recall that in list-directed or free-format input, multiple items may be separated by blank spaces, tabs, commas, or even new lines. Suppose we had tried to read the three items directly from the control file in a list-directed fashion and an item was missing from the line, Fortran would look for the missing item in the next line, making it difficult to detect exactly where the error occurred.

 Style tip

When reading noncharacter data from a file in a list-directed fashion, try reading each line into a character-string buffer first and then read from the buffer, treating it as an internal file.

5.2.4 First Snapshot of ELOGIT

We are now ready to begin coding the main ELOGIT program. Here is our first version.

```
                          ──── elogit.f90 ────
!#######################################################################
!###   ELOGIT: A simple program for logistic regression analysis  #####
!###   written entirely in standard Fortran-95.                   #####
!###   For non-PC platforms, the value of "platform" may be       #####
!###   changed to "UNIX" or "MAC", so that carriage returns are   #####
!###   handled correctly.                                         #####
!#######################################################################
program elogit
   use error_handler
   use program_constants
   use elogit_ctrlfile
   implicit none
   ! declare instances of various types
   type(error_type) :: err
   type(elogit_ctrlfile_type) :: ctrlfile
   ! additional variables and parameters for the console application
   character(len=*), parameter :: platform = "PC"
   character(len=256) :: msg_string
   character(len=file_name_length) :: control_file_name
   ! Query the user for the name of the control file
   print "(A)", "Enter name of control file:"
   read(*,"(A)") control_file_name
   print "(A)", ""
   if(control_file_name == "") then
      call err_handle(err, 1000, &
           custom_1 = "No control file name specified.")
      goto 800
   end if
   ! read control file
   if( read_elogit_ctrlfile(control_file_name, ctrlfile, err) &
        == RETURN_FAIL ) goto 800

   ! We'll add more code here later

800 continue
   ! report "OK" or error message
   if( err_msg_present(err) ) then
      call err_get_msgs(err, msg_string, platform)
      print "(A)", trim(msg_string)
      print "(A)", "Aborted"
   else
      print "(A)", "OK"
   end if
end program elogit
!#######################################################################
```

At this point, ELOGIT has four source-code files and one documentation
file. Copies of these files as they now appear are provided on our Web site
as "Snapshot 1." Readers are encouraged to examine these files, compile

them, and run the program on their own computers. An example control
file is shown below.

```
———————————————————— viral.ctl ————————————————————
**********************************************************************
* Example control file for the ELOGIT program
* Performs logistic regression on the viral assay experiment data
* from Davis et al.(1989)
**********************************************************************
* Data input section
*    LINE 1: ncase, nvar, case_id_present
*    LINE 2: data_file_name
*    LINE 3: names_file_name (if no names file provided, leave blank)
5 3 F
viral.dat
viral.nam
```

Running the program on this file causes the computer to print OK to the
screen; try introducing errors into the file to see how the program reacts.

It's natural to ask why we have taken several hundred lines of source
code to create a program that, at this point, merely reads a few items from
a text file. A novice programmer could accomplish the same task with less
than 20 lines. The benefits of a careful approach will become apparent
as the complexity of the program grows. If you adopt this style, you will
discover that your projects take shape more quickly, have fewer errors, are
much easier to maintain, and generally work better than if you cut corners
and make them shorter.

5.2.5 Exercises

1. Review the possible values of the **status** specifier for an **open** statement. Give examples where each might be useful.

2. ELOGIT prompts the user to type the name of the control file on a
   separate line. Some users may find it more convenient to provide the
   name of the control file while invoking the program, like this:

   ```
   elogit viral.ctl
   ```

   Intrinsics for retrieving command-line arguments are not part of the
   Fortran 95 standard, although they have been promised for 2003.
   Nevertheless, most compilers do supply subroutines for this purpose.
   Learn about your compiler's capabilities for handling command-line
   arguments, and modify ELOGIT to take advantage of them.

3. Write a function that counts the number of records or lines in an
   existing text file. (Hint: Use **read** statements with the **eof** specifier.)
   Add an option to ignore blank lines.

4. Write a procedure that examines a character string of arbitrary length and counts the number of words. Define a word as any amount of text delimited by blank spaces, tabs, or commas but not new lines. Use this procedure to enhance the error-reporting capabilities of the function `read_elogit_ctrlfile`.

5. Many free tools are available on the Internet for managing source code on Windows, Unix/Linux, and Macintosh computers. These are often called "version control" or "revision control" systems. Download one of these tools and learn how to use it.

5.3   Starting the Main Types Module

5.3.1   An Object-Oriented Design

Our next major task is to implement procedures for reading the data file. Before writing this code, however, let's pause to consider where we are headed. The ELOGIT console program will run in the old-fashioned batch mode. In the future, however, we may consider other styles of execution. For example, we might want to call the program from an Excel spreadsheet. We might want to embed it in a Web page or some other kind of document (e.g., an electronic textbook) to use as a teaching device. Or we might want to develop it into a Windows application with a menu-driven graphical interface. In those environments, the program would have to respond incrementally to mouse clicks and other events. It would be a shame if most of the code we were writing for the console application could not be used for these other purposes. Fortunately, by adopting an object-oriented strategy, we can package most of ELOGIT's source code as a single module whose public procedures are useful for both interactive and batch processing.

To begin this module, which we call `elogit_types`, let's start to conceptualize the object classes and methods. Suppose that we treat the entire dataset—including the numeric data values, the variable names, and the case identifiers, if present—as a single object. Creating a derived type to hold the dataset is a simple matter. We can also create additional derived types to hold the model specification, the results from the model-fitting procedure, and so on. Going further, suppose that we collect all of these derived types into a single object called an "ELOGIT session." One instance of the ELOGIT session will then store all the information about the state of the program at any time.

Storing the ELOGIT session as a single object will help us to simplify the code by keeping argument lists for procedures very short. Instead of writing one grand procedure that does everything, we will write a set of public methods to perform small, well-defined tasks, such as loading the data matrix, loading variable names, selecting the response variable, and selecting the predictors. Keeping the tasks small makes it easier to enforce rules to prevent misuse. For example, a user will not be allowed to specify any aspect of the model until the data are loaded, a variable will not be allowed as a predictor if it has already been declared as the response, and so on. A console program will invoke the methods in a predetermined order, and a violation of the rules will stop execution of the program and report an error message. In a nonconsole application, an error will not terminate the program but trigger another event (e.g., displaying a message window) that gives the user immediate feedback to correct the mistake.

Here is the beginning of our `elogit_session_type`. It is the only public type in the module. The `dataset_type`, which is private, will be defined next.

```
───────────────── elogit_types.f90 ─────────────────
!#################################################################
type :: elogit_session_type
   sequence
   private
   logical :: is_null=.true.
   type(dataset_type) :: dataset

   ! More components will be added later

end type elogit_session_type
!#################################################################
```

Notice that we included `elogit` in the name of the type; this will help to prevent name conflicts in future applications that may use the module.

Style tip───

When naming a public type, include a short prefix based on the name of the module.

5.3.2 Storing the Dataset

The next issue is to determine the types of data that the ELOGIT program will read and how the data file will be formatted. Data for logistic regression may arrive in a variety of ways, as seen by the following examples.

Grouped Responses with Numeric Covariates

The data shown in Table 5.1, reported by Davis, et al. (1989), summarize a dose-response experiment in which five groups of test animals were exposed to viral particles at various dilution levels. In this table, n_i represents the

TABLE 5.1. Data from a viral assay experiment.

Concentration	z_i	n_i	y_i
1×10^{-5}	-5	6	0
1×10^{-4}	-4	6	1
1×10^{-3}	-3	6	4
1×10^{-2}	-2	6	6
1×10^{-1}	-1	6	6

number of animals exposed in group i, and y_i is the number that died. It is of interest to determine how π_i, the probability of death, varies as a function of $z_i = \log_{10}(\text{dilution}_i)$. A reasonable model is $\text{logit}(\pi_i) = x_i^T \beta$, where $x_i = (1, z_i)$, so that $\exp(\beta_2)$ is the multiplicative increase in the odds of mortality associated with a tenfold increase in concentration.

For this example, we will suppose that the data will arrive as an ASCII file with a separate line or record for each of the five concentration groups. We will allow the user to include an arbitrary number of comment lines at the beginning of the data file, like this:

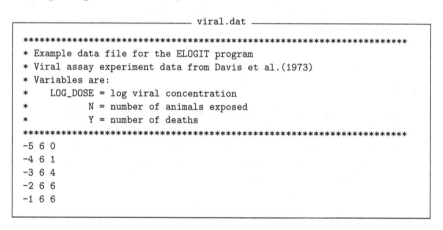

```
_____ viral.dat _____

 ****************************************************************
 * Example data file for the ELOGIT program
 * Viral assay experiment data from Davis et al.(1973)
 * Variables are:
 *     LOG_DOSE = log viral concentration
 *            N = number of animals exposed
 *            Y = number of deaths
 ****************************************************************
 -5 6 0
 -4 6 1
 -3 6 4
 -2 6 6
 -1 6 6
```

We will suppose that variable names, if provided, will be placed in a separate file with one name per line.

```
_____ viral.nam _____

 ****************************************************************
 * Example names file for the ELOGIT program
 * Variable names associated with the file viral.dat
 ****************************************************************
 LOG_DOSE
 N
 Y
```

TABLE 5.2. Results from an experiment on transient vasorestriction.

i	Volume	Rate	y_i	i	Volume	Rate	y_i
1	3.70	0.825	1	21	0.40	2.000	0
2	3.50	1.090	1	22	0.95	1.360	0
3	1.25	2.500	1	23	1.35	1.350	0
4	0.75	1.500	1	24	1.50	1.360	0
5	0.80	3.200	1	25	1.60	1.780	1
6	0.70	3.500	1	26	0.60	1.500	0
7	0.60	0.750	0	27	1.80	1.500	1
8	1.10	1.700	0	28	0.95	1.900	0
9	0.90	0.750	0	29	1.90	0.950	1
10	0.90	0.450	0	30	1.60	0.400	0
11	0.80	0.570	0	31	2.70	0.750	1
12	0.55	2.750	0	32	2.35	0.030	0
13	0.60	3.000	0	33	1.10	1.830	0
14	1.40	2.330	1	34	1.10	2.200	1
15	0.75	3.750	1	35	1.20	2.000	1
16	2.30	1.640	1	36	0.80	3.330	1
17	3.20	1.600	1	37	0.95	1.900	0
18	0.85	1.415	1	38	0.75	1.900	0
19	1.70	1.060	0	39	1.30	1.625	1
20	1.80	1.800	1				

Binary Responses with Numeric Covariates

The data shown in Table 5.2 were taken from Myers (1990, Chapter 7). $N = 39$ individuals were subjected to varying rates and volumes of air. The response is $y_i = 1$ if subject i experienced a transient vasorestriction and 0 otherwise. Because the response is binary, $n_i = 1$, $i = 1, \ldots, 39$ is understood. One model worth fitting is

$$\log\left(\frac{\pi_i}{1 - \pi_i}\right) = \beta_1 + \beta_2 \, \text{volume}_i + \beta_3 \, \text{rate}_i,$$

but we also may wish to consider simpler models based on volume or rate alone. For this example, the first few lines of the data file look like this:

```
—————————————————— vaso.dat ——————————————————
 1 3.7 .825 1
 2 3.5 1.090 1
 3 1.25 2.5 1
```

Notice that the line lengths and field widths vary; this will not be a problem if we read in the data using a list-directed format.

The first variable in this data file is merely a case identification number and will not be used in the model. Identifiers, whether they are numeric

TABLE 5.3. Graduate applicants to the University of California at Berkeley classified by sex, department, and admission status.

Dept.	Men rejected	Men accepted	Women rejected	Women accepted
A	313	512	19	89
B	207	353	8	17
C	205	120	391	202
D	278	139	244	131
E	138	53	299	94
F	351	22	317	24

or character variables, are present in many datasets and can be beneficial (e.g., when flagging influential cases or outliers). To keep our data-input routines simple, we will require the case identifier, if present, to be (a) the first variable on each line; (b) to be no longer than a certain number of characters; and (c) to have no embedded blank spaces.

Grouped Responses with Nominal Covariates

Table 5.3, reported by Freedman, Pisani and Purves (1997), describes applicants to six graduate programs at the University of California at Berkeley for the fall of 1973. This dataset is a $6\times2\times2$ contingency table with subjects classified by department, sex, and admission status. If we define "success" as being granted admission to Berkeley and regard sex and department as potential covariates, it becomes natural to group the applicants by their covariate values and rearrange the data with one covariate pattern per line, as in the data file shown below.

```
                           admissions.dat
Dept_A   Male      512    313
Dept_A   Female     89     19
Dept_B   Male      353    207
Dept_B   Female     17      8
Dept_C   Male      120    205
Dept_C   Female    202    391
Dept_D   Male      139    278
Dept_D   Female    131    244
Dept_E   Male       53    138
Dept_E   Female     94    299
Dept_F   Male       22    351
Dept_F   Female     24    317
```

This data file differs from the previous ones in two important ways. First, the responses are provided in terms of the number of successes y_i and the number of failures $n_i - y_i$ rather than in the event/trial (y_i and n_i) or binary ($y_i = 0$ or 1 with $n_i = 1$ assumed) format. Second, the two covariates in this data file are nominal rather than numeric. Nominal covariates would require us to create dummy codes or contrasts, and the interpretation of the coefficients β depends on the coding scheme. Procedures for automatically generating these dummy codes and contrasts are not difficult to write but would make our program substantially longer. To keep the project simple, we will not implement the success/failure format or nominal covariates now; these tasks are left to the interested reader as exercises.

The Dataset Type

Here is our definition for the type that will hold the dataset in the ELOGIT session. At this point, we assume that all variables are numeric. Note that arrays of unknown size must be declared as pointers (Section 3.5.1).

```
————————————— elogit_types.f90 ——————————————
!###############################################################
type :: dataset_type
   sequence
   private
   logical :: is_null=.true.
   integer(kind=our_int) :: ncase=0, nvar=0
   real(kind=our_dble), pointer :: data_matrix(:,:)=>null()
   character(len=var_name_length), pointer :: var_names(:)=>null()
   character(len=case_id_length), pointer :: case_id(:)=>null()
end type dataset_type
!###############################################################
```

This type definition relies upon two constants that have been added to the `program_constants` module:

```
integer, parameter :: var_name_length = 8, &
      case_id_length = 8
```

The next step is to create public methods for putting data in and getting data out. Before presenting these methods, however, we briefly digress to discuss the general problem of allocating array pointers.

5.3.3   A Module for Pointer Allocation

Allocating array pointers can be a dangerous operation. If the requested size of the array is too large for the available memory, the program will crash. Allocating a pointer that is currently associated will cause a memory leak. These problems can be avoided by taking two simple measures. To stop leaks, we should always check the association status of a pointer with the **associated** function and deallocate it if necessary before allocating it. Second, in every **allocate** statement, we should make use of the optional argument **stat**; this sets an integer variable to zero if the allocation was successful or to a positive value if it failed, so that errors can be trapped.

If pointers are frequently allocated in a program, these checks can become tedious. To avoid unnecessary duplication of code, we have written a module called `dynamic_allocation` that implements these checks and reports violations to the error handler. To see how the module works, consider the following procedure for allocating a rank-two integer array.

```
————————————————————— dynalloc.f90 —————————
!####################################################################
integer(our_int) function int2_alloc(intArray, dim1, dim2, err) &
    result(answer)
  ! Allocates an integer array of rank 2
  implicit none
  ! declare required arguments
  integer(kind=our_int), pointer :: intArray(:,:)
  integer, intent(in) :: dim1, dim2
  type(error_type), intent(inout) :: err
  ! declare local variables and parameters
  integer :: status
  character(len=*), parameter :: subname = "int2_alloc"
  ! begin
  answer = RETURN_FAIL
  if( associated(intArray) ) deallocate(intArray, stat=status)
  if( status /= 0 ) goto 800
  allocate( intArray(dim1, dim2), stat=status )
  if(status /= 0 ) goto 810
  ! normal exit
  answer = RETURN_SUCCESS
  return
  ! error traps
800   call err_handle(err, 201, &
          called_from = subname//" in MOD "//modname)
  return
810   call err_handle(err, 200, &
          called_from = subname//" in MOD "//modname)
  return
end function int2_alloc
!####################################################################
```

The module has similar functions for allocating integer arrays of other ranks, along with arrays of real numbers and strings. To avoid having to remember different names for each of these functions, we group them together under the generic name `dyn_alloc` and allow Fortran to select the appropriate procedure to apply based on the type and number of arguments. The module also has a parallel set of procedures for deallocating arrays, grouped under the generic name `dyn_dealloc`.

Using all of this code to handle the occasional allocation error may seem a bit extreme. However, you don't need to rewrite this code for your own programs. If you download `dynalloc.f90` and include the statement

```
use dynamic_allocation
```

in your program or module, you can manage pointer allocation exactly as
we do.

5.3.4 Putting and Getting Data

Returning now to the problem at hand, let us consider how to put a dataset
into the ELOGIT session. We are working toward a procedure that reads
data values from a file. But first, we should write put methods that accept
data not from external files but from arrays of real numbers and strings. The
reason is that we want to enable the elogit_types module to serve other
programs written in Fortran or perhaps in other languages. We do not want
file I/O to be the only method by which ELOGIT may communicate with
the outside world. Rather, we want the option of bypassing files entirely
to insert the contents of arrays or array sections directly into the ELOGIT
session.

Referring back to our definition of dataset_type, we see that the most
crucial component is data_matrix, a rank-two real array for storing the
numeric values for all the variables in the system. The integer components
ncase and nvar are merely the dimensions of data_matrix, the logical flag
is_null merely tells whether data_matrix is present, and the character
arrays var_names and case_id merely provide names for its rows and
columns. Therefore, we will write the put method for the data_matrix first.
As a side effect, this method will reset the values of is_null, ncase, and
nvar. It will also allocate var_names and case_id to the correct length and
fill them with sensible default strings. In this way, var_names and case_id
will exist whether or not a user chooses to supply them. Our put method
for the data matrix is shown below.

```
                        elogit_puts.f90
!####################################################################
integer(our_int) function put_elogit_data_matrix(data_matrix, &
     session, err) result(answer)
! Loads a data matrix into an ELOGIT session.
! When the data matrix is loaded, var_names and case_id are
! initialized to default values. If the loading fails for any
! reason, the entire ELOGIT session is nullified.
implicit none
! declare arguments
real(kind=our_dble), pointer :: data_matrix(:,:)
type(elogit_session_type), intent(inout) :: session
type(error_type), intent(inout) :: err
! declare local variables and parameters
integer(kind=our_int) :: ijunk
character(len=*), parameter :: subname = &
     "put_elogit_data_matrix"
! check arguments
answer = RETURN_FAIL
if( .not.associated(data_matrix) ) goto 700
! nullify the entire session
```

```
         if( nullify_elogit_session(session, err) == RETURN_FAIL ) &
            goto 800
         ! transfer dimensions to dataset, allocate pointers in dataset
         session%dataset%is_null = .false.
         session%dataset%ncase = size(data_matrix, 1)
         session%dataset%nvar = size(data_matrix, 2)
         if( dyn_alloc(session%dataset%data_matrix, &
            session%dataset%ncase, session%dataset%nvar, err) &
            == RETURN_FAIL ) goto 800
         if( dyn_alloc(session%dataset%var_names, &
            session%dataset%nvar, err) == RETURN_FAIL ) goto 800
         if( dyn_alloc(session%dataset%case_id, &
            session%dataset%ncase, err) == RETURN_FAIL ) goto 800
         ! transfer contents from data_matrix to dataset%data_matrix
         session%dataset%data_matrix = data_matrix
         ! initialize var_names and case_id
         call assign_default_var_names(session%dataset)
         call assign_default_case_id(session%dataset)
         ! normal exit
         answer = RETURN_SUCCESS
         return
         ! error traps
 700     call err_handle(err, 1000, &
            called_from = subname//" in MOD "//modname, &
            custom_1 = "Input array is null.")
         ijunk = nullify_elogit_session(session, err)
         return
 800     call err_handle(err, 1000, &
            called_from = subname//" in MOD "//modname)
         ijunk = nullify_elogit_session(session, err)
      end function put_elogit_data_matrix
      !####################################################################
```

This procedure relies on a private subroutine for initializing the variable
names, and a similar procedure (not shown) for initializing the case identifiers.

```
_____ elogit_puts.f90 _____
      !####################################################################
      subroutine assign_default_var_names(dataset)
         ! assigns default values to dataset%var_names, assuming that
         ! the array has already been allocated
         implicit none
         ! declare arguments
         type(dataset_type), intent(inout) :: dataset
         ! declare local variables and parameters
         character(len=12) :: sInt
         integer(kind=our_int) :: var
         ! begin
         do var = 1, dataset%nvar
            write(sInt,"(I12)") var
            sInt = adjustl(sInt)
            dataset%var_names(var) = "VAR_" // trim(sInt)
```

```
      end do
   end subroutine assign_default_var_names
   !################################################################
```

To keep the source-code file for the `elogit_types` module from getting too large, we are placing this function and all other puts related to the ELOGIT session into an auxiliary file called `elogit_puts.f90`. We have then added the line

```
   include "elogit_puts.f90"
```

to `elogit_types.f90`, which instructs the compiler to insert all the statements from `elogit_puts.f90` at that point. Similarly, we will be placing the get methods into a file called `elogit_gets.f90`. If you try to compile this project automatically with an IDE or make file, make sure that you *do not* include these auxiliary files in the compilation list; otherwise the compiler will try to process them twice.

Style tip

As a module grows in size, place groups of procedures into auxiliary files and use `include` statements in the original file.

Now we write put methods for the other properties. There is no need for public methods for putting `ncase`, `nvar`, or `is_null`, as these properties should be read-only. Here is a public procedure for putting `var_names`; a similar procedure for putting `case_id` is not shown.

```
──────────────────────── elogit_puts.f90 ───────────
!################################################################
integer(our_int) function put_elogit_var_names(var_names, &
     session, err) result(answer)
   ! Loads an array of variable name strings into an ELOGIT
   ! session after a data matrix has already been loaded.
   ! Side effect: the rest of the ELOGIT session (everything after
   ! the model) will be nullified.
   implicit none
   ! declare arguments
   character(len=var_name_length), pointer :: var_names(:)
   type(elogit_session_type), intent(inout) :: session
   type(error_type), intent(inout) :: err
   ! declare local variables and parameters
   integer :: ijunk
   character(len=*), parameter :: subname = "put_elogit_var_names"
   ! check arguments
   answer = RETURN_FAIL
   if( session%dataset%is_null ) goto 700
   if( .not.associated(var_names) ) goto 800
   if( size(var_names) /= session%dataset%nvar ) goto 810
   ! transfer contents from var_names to dataset%var_names
```

```
        session%dataset%var_names = var_names
        ! reset the other parts of the session
        ijunk = nullify_elogit_session( session, err, &
            save_dataset = .true. )
        ! normal exit
        answer = RETURN_SUCCESS
        return
        ! error traps
700     call err_handle(err, 1000, &
            called_from = subname//" in MOD "//modname, &
            custom_1 = "You cannot load var_name strings", &
            custom_2 = "until a data matrix has been loaded.")
        return
800     call err_handle(err, 1000, &
            called_from = subname//" in MOD "//modname, &
            custom_1 = "Input array is null.")
        return
810     call err_handle(err, 1000, &
            called_from = subname//" in MOD "//modname, &
            custom_1 = "Size of input array does not conform", &
            custom_2 = "to the current dataset.")
    end function put_elogit_var_names
    !#################################################################
```

A side effect of this method is that all aspects of the ELOGIT session other than the dataset are nullified. At the moment, the session has no other aspects. However, we will soon add components to hold the model specification, the results from the model-fitting procedures, and so on. Some of those session properties (e.g., the model specification) could become irrelevant if the variable names are changed. For that reason, we automatically erase those components whenever the dataset is changed through a call to nullify_elogit_session.

Now that the puts are finished, let's implement the gets. Gets are not always necessary, and some of them may not be called in the ELOGIT console program. They may be needed in the future, however, if the elogit_types module is used by other projects. Shown below is the procedure for getting variable names; the other gets follow the same pattern and can be easily generated by making minor changes to this one.

```
———————————————— elogit_gets.f90 ————————————————
    !#################################################################
    integer(our_int) function get_elogit_var_names(var_names, &
        session, err) result(answer)
        ! Gets the var_names currently stored in an ELOGIT session
        implicit none
        ! declare arguments
        character(len=var_name_length), pointer :: var_names(:)
        type(elogit_session_type), intent(in) :: session
        type(error_type), intent(inout) :: err
        ! declare local variables and parameters
```

```
      character(len=*), parameter :: subname = "get_elogit_var_names"
      ! begin
      answer = RETURN_FAIL
      if( session%dataset%is_null ) goto 700
      if( dyn_alloc(var_names, session%dataset%nvar, err) &
          == RETURN_FAIL ) goto 800
      var_names = session%dataset%var_names
      ! normal exit
      answer = RETURN_SUCCESS
      return
      ! error traps
  700 call err_handle(err, 1000, &
          called_from = subname//" in MOD "//modname, &
          custom_1 = "No dataset has been loaded yet.")
      return
  800 call err_handle(err, 1000, &
          called_from = subname//" in MOD "//modname)
    end function get_elogit_var_names
    !###############################################################
```

You may have noticed that the array arguments in these puts and gets have the pointer attribute. This will make it easier for other program units to use the methods. If we had not done this, then a program that called `get_elogit_var_names` would have to ascertain the size of `var_names` in advance and allocate an array of that size before calling the method. Declaring the array arguments as pointers allows us to perform all dimensioning inside of the module procedures. Pointers will also facilitate the process of turning your program into a COM server in the future. When writing puts and gets for scalars, the pointer attribute is unnecessary.

 Style tip

When writing a public method for putting or getting an array, declare the array as a pointer. Do not use the pointer attribute for scalars.

Is it really necessary to write these puts and gets? If our only purpose is to develop a single working console program, perhaps not. However, if we intend to use this module in future projects, there are tremendous benefits to having a self-contained, encapsulated object, complete with error handling. We would simply need to put the line `use elogit_types` into our code; immediately, all of the puts, gets, and computational methods are available. In addition, we may wish to move beyond console applications to the interactive, point-and-click world of Windows. Procedure calls will no longer occur in a prespecified sequence, and the possibilities for erroneous input to the module grow exponentially. In this setting, anticipating all the ways that users can misapply the procedures and preventing them from doing so becomes crucial. In Windows, it is the quality of the interface—

i.e., the entire set of public procedures, including the puts and gets—that primarily determines how robust and useful the `elogit_types` module will be.

5.3.5 Reading Data from Files

We are finally ready to create a routine for reading data files. We have written a public function called `read_elogit_datafile` that:

- allocates a temporary rank-two real array to hold the data matrix and, if necessary, a temporary rank-one string array to hold case identifiers;

- opens the data file and skips any comment lines at the beginning;

- reads in each line of data from the file, storing the case identifier (if present) and the numeric data in the temporary arrays;

- puts the data matrix into `dataset` using `put_elogit_data_matrix`; and

- puts the case identifiers, if present, into `dataset` using the procedure `put_elogit_case_id`.

Any case identifier that exceeds the maximum number of characters is truncated, and a warning is generated. The warning message is stored in a second instance of the `error_type`.

```
_____ elogit_io.f90 _____

!#################################################################
integer(kind=our_int) function read_elogit_datafile( &
    data_file_name, nkase, nvar, case_id_present, session, err, &
    warn) result(answer)
! Reads a data matrix from a data file and stores it in an
! ELOGIT session. If the operation fails for any reason, the
! session is nullified and the returned value is RETURN_FAIL.
! Generates an optional warning message if case_id strings had
! to be truncated.
implicit none
! declare required arguments
character(len=file_name_length), intent(in) :: data_file_name
integer(kind=our_int), intent(in) :: nkase, nvar
logical, intent(in) :: case_id_present
type(elogit_session_type), intent(inout) :: session
type(error_type), intent(inout) :: err
! declare optional arguments
type(error_type), intent(inout), optional :: warn
! declare local variables and parameters
real(kind=our_dble), pointer :: data_matrix(:,:)
character(len=case_id_length), pointer :: case_id(:)
character(len=data_line_width) :: line
```

```fortran
integer(kind=our_int) :: kase, var, posn, current_line, ijunk
logical :: truncate_warn
character(len=*), parameter :: subname = "read_elogit_datafile"
! check the arguments
answer = RETURN_FAIL
if(data_file_name == "") goto 700
if(nkase <= 0) goto 710
if(nvar <= 0) goto 720
! allocate data_matrix and case_id, if necessary
if( dyn_alloc(data_matrix, nkase, nvar, err) == RETURN_FAIL ) &
     goto 800
if( case_id_present ) then
   if( dyn_alloc(case_id, nkase, err) == RETURN_FAIL ) goto 800
end if
! open the data file and ignore comment lines
open(unit=data_file_handle, file=data_file_name, &
     status="old", err=810)
current_line = 0
if( skip_comment_lines(data_file_handle, current_line) &
     == RETURN_FAIL ) goto 900
! read data
truncate_warn = .false.
kase = 0
do
   current_line = current_line + 1
   read(unit=data_file_handle, fmt="(A)", err=900, end=900) line
   if(line == "") goto 900    ! no blank lines allowed
   kase = kase + 1
   if( case_id_present ) then
      line = adjustl(line)
      posn = index(line, " ") - 1 ! length of current case_id
      if( posn > case_id_length ) truncate_warn = .true.
      case_id(kase) = line(1:posn)
      line = line(posn + 1:)          ! remove case_id from line
   end if
   read(line, *, err=900, end=900) &
        ( data_matrix(kase, var),  var = 1, nvar )
   if(kase == nkase) exit
end do
! load the data matrix
if( put_elogit_data_matrix(data_matrix, session, err) == &
     RETURN_FAIL ) goto 800
! load the case_id, if present
if( case_id_present ) then
   if( put_elogit_case_id(case_id, session, err) == &
        RETURN_FAIL ) goto 800
end if
! issue warnings, if warranted
if( present(warn) ) then
   if( truncate_warn ) &
        call err_handle( warn, 1000, &
        called_from = subname//" in MOD "//modname, &
        custom_1 = &
```

```
                    "One or more case identifiers were truncated.")
        end if
        ! normal exit
        answer = RETURN_SUCCESS
        goto 999
        ! error traps
700     call err_handle(err, 1000, &
            called_from = subname//" in MOD "//modname, &
            custom_1 = "No data file name specified.")
        goto 999
710     call err_handle(err, 1000, &
            called_from = subname//" in MOD "//modname, &
            custom_1 = "Number of cases not positive.")
        goto 999
720     call err_handle(err, 1000, &
            called_from = subname//" in MOD "//modname, &
            custom_1 = "Number of variables not positive.")
        goto 999
800     call err_handle(err, 1000, &
            called_from = subname//" in MOD "//modname)
        goto 999
810     call err_handle(err, 1, &
            called_from = subname//" in MOD "//modname, &
            file_name = data_file_name)
        goto 999
900     call err_handle(err, 3, &
            called_from = subname//" in MOD "//modname, &
            file_name = data_file_name, line_no = current_line)
        goto 999
        ! final cleanup
999     continue
        close(unit=data_file_handle)
        if( answer == RETURN_FAIL ) &
            ijunk = nullify_elogit_session(session, err)
        ijunk = dyn_dealloc(data_matrix, err)
        ijunk = dyn_dealloc(case_id, err)
    end function read_elogit_datafile
    !##############################################################
```

Another function called `read_elogit_namesfile`, which is not shown, performs a similar operation on the variable names file.

5.3.6 Second Snapshot of ELOGIT

Implementation of the data-reading procedures represents an important milestone in the development of ELOGIT. This is an excellent time to stop and test these new procedures thoroughly. To test them, we need to add a few lines to the main program after the reading of the control file.

```
─────────────── elogit.f90 ───────────────
    ! read the data file
    if( read_elogit_datafile(ctrlfile%data_file_name, &
```

```
        ctrlfile%ncase, ctrlfile%nvar, ctrlfile%case_id_present, &
        session, err, warn) == RETURN_FAIL ) goto 800
    ! read the variable names file, if present
    if( ctrlfile%names_file_present ) then
      if( read_elogit_namesfile(ctrlfile%names_file_name, &
          ctrlfile%nvar, session, err, warn) == RETURN_FAIL ) &
          goto 800
    end if
```

An image of ELOGIT as it now stands is provided on our Web site as "Snapshot 2." Download the snapshot and compile the source files. Then try to run the program using the data file **viral.dat** and names file **viral.nam** from Section 5.3.2.

As part of Snapshot 2, we have produced a short document for users of ELOGIT that clearly states the requirements for the data and names file. The document is not fancy, but it is complete and precise.

```
——————————— datafilespec.txt ———————————

FORMAT OF AN ELOGIT DATA FILE
-----------------------------

A rectangular dataset for the ELOGIT program is provided through an
ASCII file. Case identifier strings are optional.

COMMENT LINES: An unlimited number of comment lines is allowed at the
beginning of the data file. Each comment line must begin with an
asterisk "*". These comment lines are ignored.

IF NO CASE IDENTIFIER STRINGS ARE PRESENT, then the lines after the
comments should look like this:

LINE 1:     Var_1 Var_2   ...   Var_nvar
LINE 2:     Var_1 Var_2   ...   Var_nvar
    .
    .
    .
LINE ncase: Var_1 Var_2   ...   Var_nvar

The variables on each line may be separated by any amount of blank
space. Each line must have all variables on it; blank lines between
cases are not allowed.

IF CASE IDENTIFIER STRINGS ARE PRESENT, then the lines after the
comments should look like this:

LINE 1:     Case_id Var_1 Var_2   ...   Var_nvar
LINE 2:     Case_id Var_1 Var_2   ...   Var_nvar
    .
    .
    .
LINE ncase: Case_id Var_1 Var_2   ...   Var_nvar

The case_id is a character string up to 8 characters long, with no
```

> embedded spaces. A case_id string should not begin with an asterisk
> "*", because then it might be mistaken for a comment. Blank spaces on
> a line before the case_id string are ignored.

A similar document (not shown) explains the requirements for the optional variable names file. It's tempting to put off writing these documents until later, after the entire program is working. If you wait, however, they will take longer to write and may be less accurate.

 Style tip ───

The best window of opportunity for documenting file requirements is while you are writing the procedures for reading those files.

───

5.3.7 Exercises

1. Our data input routines for ELOGIT require the user to specify the number of variables and number of cases in the data file. Revise them to determine the number of variables and the number of cases automatically. Note that this will also require revision of the control file.

2. In Section 3.1.5, we described an object class for performing the classical chi-square test for independence in a two-way contingency table. Implement this object class as a derived type. Place the type in a module, and write all the necessary put and get methods. (Don't implement the `run_chisquare_test` method yet.) Can you think of any more properties that would be useful?

5.4 Specifying the Model

5.4.1 Storing the Model Specification

The ELOGIT session now accepts data, but it does not yet know any details of the model to be fit. Here is a derived type that can hold those details.

```
─────────────── elogit_types.f90 ───────────────
!##################################################################
type :: model_type
   sequence
   private
   logical :: is_null=.true.
   integer(kind=our_int) :: y_col=0, n_col=0, npred=0
   logical :: grouped=.false., intercept_present=.true.
   integer(kind=our_int), pointer :: pred_col(:)=>null()
```

```
end type model_type
!###################################################################
```

The component `y_col` is an integer that specifies which column of the data matrix contains the response y_i; `n_col` specifies which column, if any, contains n_i; `grouped` is `.false.` if the responses are binary (i.e., if all n_i's are assumed to be one) and `.true.` if n_i's are explicitly provided; `pred_col` is a rank-one array of integers specifying which columns of the data matrix contain the covariates; `npred` is the number of covariates (i.e., the size of `pred_col`); and `intercept_present` indicates whether an intercept is to be included in the model. The `is_null` component, which is initially `.true.`, will be set to `.false.` once a valid model has been specified.

Before proceeding, we make a distinction between the properties of the model, which are conceptual entities, and the components of this derived type, which are Fortran scalars and arrays. There is not a one-to-one correspondence between them. The properties of the model are (a) what the response is, (b) what the predictors are, and (c) whether an intercept is to be included. The details of how these properties are stored in the `elogit_session_type` are irrelevant to the user and to any program unit outside of the `elogit_types` module.

5.4.2 Putting and Getting Model Properties

Our first put method, `put_elogit_response_bycol`, identifies the response variable. The input argument `col` points to an array of integers. If the response is binary, this array should contain a single integer indicating which column of the dataset contains y_i. If the response is binomial, then the array should have two elements—one specifying the column for y_i, the other specifying the column for n_i. Notice that the argument list includes an `elogit_dataset_type` as well as an `elogit_model_type`, enabling us to perform some intelligent consistency checks. First, we make sure that a dataset has been loaded. Next, we make sure that the variables exist (i.e., that the column numbers for y_i and n_i lie between 1 and `dataset%nvar`). Finally, we make sure that $0 \le y_i \le n_i$ for all cases. If no violations occur, then the model components `y_col`, `n_col`, and `grouped` are set to the desired values. As a side effect, this method also sets the values of `npred`, `pred_col`, and `intercept_present` to indicate a null model with an intercept but no other predictors. Predictors will be introduced later by additional put methods.

```
——————————————————— elogit_puts.f90 ———————————————————
!###################################################################
integer(our_int) function put_elogit_response_bycol(col, session, &
     err) result(answer)
  ! Declares which column in the dataset contains the response or
  ! y variable, and which column (if any) contains the n.
```

```
!
! The input argument col is a pointer to an integer array.
! For ungrouped data, the array should have size 1; the only
! element should be the column number of the y-variable.
! For grouped data, the size should be 2, and the elements
! should be the column number for the y-variable, followed by
! the column number for the n-variable.
!
! Also checks the dataset to make sure that all the y and n
! values are valid: 0.<=y<=1. for grouped data and 0.<=y<=n
! for ungrouped data.
!
! Calling this procedure has the effect of resetting the
! predictors to a null (intercept-only) model.
!
! Another side effect is that everything in the ELOGIT
! session after the data and model is nullified.
!
! If this procedure fails for any reason, the model is
! nullified.
implicit none
! declare arguments
integer(kind=our_int), pointer :: col(:)
type(elogit_session_type), intent(inout) :: session
type(error_type), intent(inout) :: err
! declare local variables and parameters
character(len=*), parameter :: subname = &
     "put_elogit_response_bycol"
integer(kind=our_int) :: y, n, kase, ijunk
real(kind=our_dble) :: ytmp, ntmp
character(len=12) :: sInt
! check arguments and set the y and n variables
answer = RETURN_FAIL
if( session%dataset%is_null ) goto 700
if( .not.associated(col) ) goto 710
if( size(col)==1 ) then
    y = col(1)
    if( (y <= 0) .or. (y > session%dataset%nvar) ) goto 740
    session%model%y_col = y
    session%model%grouped = .false.
else if( size(col)==2 ) then
    y = col(1)
    n = col(2)
    if( (n <= 0) .or. (n > session%dataset%nvar) .or. &
        (n == y) ) goto 750
    session%model%y_col = y
    session%model%n_col = n
    session%model%grouped = .true.
else
    goto 760
end if
! set the predictors to a null (intercept-only) model
session%model%intercept_present = .true.
```

```
      session%model%npred = 0
      if( dyn_dealloc(session%model%pred_col, err) &
          == RETURN_FAIL ) goto 800
      ! check the data for y and n to make sure it looks okay
      do kase = 1, session%dataset%ncase
          write(sInt, "(I12)") kase
          sInt = adjustl(sInt)
          ytmp = session%dataset%data_matrix(kase, &
              session%model%y_col)
          if( session%model%grouped ) then
              ntmp = session%dataset%data_matrix(kase, &
                  session%model%n_col)
              if( ntmp < 0.D0 ) goto 810
              if( (ytmp < 0.D0) .or. (ytmp > ntmp) ) goto 820
          else
              if( (ytmp < 0.D0) .or. (ytmp > 1.D0) ) goto 820
          end if
      end do
      session%model%is_null = .false.
      ! nullify everything in the session after the data and model
      ijunk = nullify_elogit_session( session, err, &
          save_dataset = .true., save_model = .true. )
      ! normal exit
      answer = RETURN_SUCCESS
      return
      ! error traps
700   call err_handle(err, 1000, &
          called_from = subname//" in MOD "//modname, &
          custom_1 = "You cannot specify a model", &
          custom_2 = "until a dataset has been loaded.")
      goto 999
710   call err_handle(err, 1000, &
          called_from = subname//" in MOD "//modname, &
          custom_1 = "Input array is null")
      goto 999
740   call err_handle(err, 1000, &
          called_from = subname//" in MOD "//modname, &
          custom_1 = &
          "Invalid column number given for the y-variable")
      goto 999
750   call err_handle(err, 1000, &
          called_from = subname//" in MOD "//modname, &
          custom_1 = &
          "Invalid column number given for the n-variable")
      goto 999
760   call err_handle(err, 1000, &
          called_from = subname//" in MOD "//modname, &
          custom_1 = "Invalid size for input array")
      goto 999
800   call err_handle(err, 1000, &
          called_from = subname//" in MOD "//modname)
      goto 999
810   call err_handle(err, 1000, &
```

```
           called_from = subname//" in MOD "//modname, &
           custom_1 = "Invalid data value for the n-variable", &
           custom_2 = "Case number = " // trim(sInt), &
           custom_3 = "Case id = " // &
           trim( session%dataset%case_id(kase) ) )
      goto 999
820   call err_handle(err, 1000, &
           called_from = subname//" in MOD "//modname, &
           custom_1 = "Invalid data value for the y-variable", &
           custom_2 = "Case number = " // trim(sInt), &
           custom_3 = "Case id = " // &
           trim( session%dataset%case_id(kase) ) )
      goto 999
      ! cleanup if an error occurs
999   continue
      ijunk = nullify_elogit_session( session, err, &
           save_dataset = .true.)
   end function put_elogit_response_bycol
   !###############################################################
```

From the user's standpoint, it may be more convenient to identify the y_i and n_i variables not by column number but by name. For this reason, we wrote another procedure, called put_elogit_response_byname (not shown), in which the input argument is a pointer to an array of character strings containing the variable names for y_i and, if applicable, n_i. We also added the following interface block, which allows both procedures to be invoked by the same generic name.

```
─────────────────── elogit_types.f90 ───────────────────
   interface put_elogit_response
      module procedure put_elogit_response_bycol
      module procedure put_elogit_response_byname
   end interface
```

Corresponding to these puts, we also created a pair of gets that retrieve the column numbers and names of the y_i and n_i variables. These two functions are invoked by the generic name get_elogit_response.

Now we need puts and gets for the predictors. Predictors may be any variables in the dataset other than y_i and n_i. Most users will want to include a constant term for an intercept, but we should also allow them to remove the constant if desired. Therefore, we have written:

- put_elogit_pred_bycol, to declare which columns of the dataset contain the predictors,

- put_elogit_pred_byname, which declares predictors by name rather than column number;

- put_elogit_intercept, which introduces or removes the intercept; and

  • a get procedure corresponding to each of these puts.

Once again, `put_elogit_pred_bycol` and `put_elogit_pred_byname` are grouped together by an interface block so that other program units can refer to them by the generic public name `put_elogit_pred`. Source code for all of these puts and gets can be found in `elogit_model.f90`.

5.4.3 Third Snapshot

Now that we have created an object class for the model and established its properties, it's time to add the model specification to our control file-reading procedures. We do this in four steps. First, we design a new section for the control file and describe it in our official document:

```
──────────────────── controlfilespec.txt ────────────────────
*************************************************************************
* Model specification section
* Optional comment lines, followed by:
LINE 1: by_name, grouped
        by_name = T if the variables will be specified by name,
            F if variables will be specified by number
        grouped = T if an n-variable is present, F otherwise
LINE 2: yvar, nvar
        yvar = name or number of the response variable
        nvar = name or number of the n-variable, if any (if none,
            leave this blank)
LINE 3: intercept_present
        T if the model should contain an intercept, F otherwise
LINE 4: npred
        number of predictors in the model (could be 0); predictors
        do not include the intercept if present
LINE 5: name or number of the first predictor
LINE 6: name or number of the second predictor
    and so on...
```

Second, we expand `elogit_ctrlfile_type` to hold the new information:

```
──────────────────── elogit_ctrlfile.f90 ────────────────────
!#################################################################
type :: elogit_ctrlfile_type
  ! unlike other types, the contents of this one are public
  sequence
  ! data input section
  integer(kind=our_int) :: ncase=0, nvar=0
  logical :: case_id_present=.false., names_file_present=.false.
  character(len=file_name_length) :: data_file_name="", &
      names_file_name=""
  ! model specification section
  logical :: by_name=.false., grouped=.false.
  integer(kind=our_int), pointer :: resp_col(:)=>null()
  character(len=var_name_length), pointer :: resp_name(:)=>null()
```

```
      logical :: intercept_present=.false.
      integer(kind=our_int) :: npred=0
      integer(kind=our_int), pointer :: pred_col(:)=>null()
      character(len=var_name_length), pointer :: &
          pred_names(:)=>null()

      ! We'll add more components to this type later

   end type elogit_ctrlfile_type
   !###############################################################
```

Third, we add a new section to the `read_elogit_ctrlfile` procedure:

```
                        ──── elogit_ctrlfile.f90 ────
      !##############################
      ! Model specification section
      if( skip_comment_lines(ctrl_file_handle, current_line) &
          == RETURN_FAIL ) goto 900
      ! read by_name, grouped
      current_line = current_line + 1
      read(unit=ctrl_file_handle, fmt="(A)", err=900, end=900) line
      read(line, *, err=900, end=900) ctrlfile%by_name, &
          ctrlfile%grouped
      ! read yvar, nvar
      current_line = current_line + 1
      read(unit=ctrl_file_handle, fmt="(A)", err=900, end=900) line
      if( ctrlfile%by_name ) then
         if( ctrlfile%grouped ) then
            if( dyn_alloc(ctrlfile%resp_name, 2, err) == &
                RETURN_FAIL ) goto 800
         else
            if( dyn_alloc(ctrlfile%resp_name, 1, err) == &
                RETURN_FAIL ) goto 800
         end if
         ! set pred_name
         if( line == "" ) goto 900
         line = adjustl(line)
         posn = index(line, " ")
         ctrlfile%resp_name(1) = line(:posn-1)
         line = line(posn:)
         if( ctrlfile%grouped ) then
            if( line == "" ) goto 900
            line = adjustl(line)
            posn = index(line, " ")
            ctrlfile%resp_name(2) = line(:posn-1)
         end if
      else
         ! set pred_col
         if( ctrlfile%grouped ) then
            if( dyn_alloc(ctrlfile%resp_col, 2, err) == &
                RETURN_FAIL ) goto 800
            read(line, *, err=900, end=900) ctrlfile%resp_col(1), &
                ctrlfile%resp_col(2)
```

```
          else
              if( dyn_alloc(ctrlfile%resp_col, 1, err) == &
                  RETURN_FAIL ) goto 800
              read(line, *, err=900, end=900) ctrlfile%resp_col(1)
          end if
      end if
      ! read intercept_present
      current_line = current_line + 1
      read(unit=ctrl_file_handle, fmt="(A)", err=900, end=900) line
      read(line, *, err=900, end=900) ctrlfile%intercept_present
      ! read npred
      current_line = current_line + 1
      read(unit=ctrl_file_handle, fmt="(A)", err=900, end=900) line
      read(line, *, err=900, end=900) ctrlfile%npred
      ! read predictor variables, if any
      if( ctrlfile%npred > 0 ) then
          if( ctrlfile%by_name ) then
              if( dyn_alloc(ctrlfile%pred_names, ctrlfile%npred, err) &
                  == RETURN_FAIL ) goto 800
              do i = 1, ctrlfile%npred
                  current_line = current_line + 1
                  read(unit=ctrl_file_handle, fmt="(A)", err=900, &
                      end=900) line
                  if( line == "" ) goto 900
                  line = adjustl(line)
                  if( line(1:1) == "*" ) goto 900
                  posn = index(line, " ")
                  ctrlfile%pred_names(i) = line(:posn-1)
              end do
          else
              if( dyn_alloc(ctrlfile%pred_col, ctrlfile%npred, err) &
                  == RETURN_FAIL ) goto 800
              do i = 1, ctrlfile%npred
                  current_line = current_line + 1
                  read(unit=ctrl_file_handle, fmt="(A)", err=900, &
                      end=900) line
                  read(line, *, err=900, end=900) ctrlfile%pred_col(i)
              end do
          end if
      end if
```

Finally, we add a few lines to the main program, which, after the control file has been read, puts the model properties into the model object:

```
─────────── elogit.f90 ───────────
    ! specify the model
    if( ctrlfile%by_name ) then
        if( put_elogit_response( ctrlfile%resp_name, session, &
            err) == RETURN_FAIL ) goto 800
        if( put_elogit_intercept( ctrlfile%intercept_present, session, &
            err) == RETURN_FAIL ) goto 800
        if( put_elogit_predictors( ctrlfile%pred_names, session, &
            err) == RETURN_FAIL ) goto 800
```

```
else
   if( put_elogit_response( ctrlfile%resp_col, session, &
       err) == RETURN_FAIL ) goto 800
   if( put_elogit_intercept( ctrlfile%intercept_present, session, &
       err) == RETURN_FAIL ) goto 800
   if( put_elogit_predictors( ctrlfile%pred_col, session, &
       err) == RETURN_FAIL ) goto 800
end if
```

The current state of the program is saved as "Snapshot 3." You may find it helpful to download this set of files, compile them, and test the new model specification routines.

5.4.4   Exercises

1. Our development of the `elogit_model_type` supposes that a dataset will be described by a single model. One can also imagine situations where multiple models would be applied to the same dataset. Outline a programming strategy for applying multiple models to one dataset. (Hint: Consider a linked list.)

2. Write a public function called `add_elogit_pred` for adding one or more predictor variables to an existing ELOGIT model. Overload this function so that the additional predictors may be specified either by column or by name. Write a companion function for removing predictors from an existing model.

3. Popular statistical packages, including SAS and S-PLUS, allow the user to specify a regression model through a character-string formula. A logistic model for binary data might be written as

   ```
   y ~ x1 + x2
   ```

   where `y` is the binary response and `x1` and `x2` are predictors. For a grouped response, it might look like

   ```
   y / n ~ x1 + x2
   ```

   where `y` is the number of successes and `n` is the number of trials. Develop a method that allows you to specify an ELOGIT model as a formula.

5.5   Fitting the Model

5.5.1   The Computational Task

Now that the model specification routines are finished, we set up the procedures for estimating the coefficients. Recall that the model is $y_i \sim \text{Bin}(n_i, \pi_i)$, $i = 1, \ldots, N$, where

$$\text{logit}(\pi_i) = \log\left(\frac{\pi_i}{1 - \pi_i}\right) = x_i^T \beta.$$

Solving for π_i gives

$$\pi_i = \text{expit}(x_i^T \beta) = \frac{e^{x_i^T \beta}}{1 + e^{x_i^T \beta}}.$$

Taking the logarithm of the probability mass function

$$f(y_i) = \frac{n_i!}{y_i!(n_i - y_i)!} \, \pi_i^{y_i} \, (1 - \pi_i)^{n_i - y_i}$$

and summing over the cases produces the loglikelihood function. Omitting terms that do not involve parameters, the loglikelihood is

$$
\begin{aligned}
l(\beta) &= \sum_{i=1}^{N} \{ y_i \log \pi_i + (n_i - y_i) \log(1 - \pi_i) \} \\
&= \sum_{i=1}^{N} y_i \log\left(\frac{\pi_i}{1 - \pi_i}\right) - \sum_{i=1}^{N} n_i \log\left(\frac{1}{1 - \pi_i}\right) \\
&= \sum_{i=1}^{N} (x_i^T \beta) \, y_i - \sum_{i=1}^{N} n_i \log\left(1 + e^{x_i^T \beta}\right).
\end{aligned}
\tag{5.2}
$$

Except in trivial cases, no closed-form expression exists for the maximizer of $l(\beta)$, so we must find it by an iterative method.

5.5.2   Newton-Raphson and Weighted Least Squares

The traditional way to compute the maximum-likelihood estimate for $\beta = (\beta_1, \ldots, \beta_p)^T$ in the logistic model is by the Newton-Raphson procedure. Newton-Raphson updates the current estimate $\beta^{(t)}$ by

$$\beta^{(t+1)} = \beta^{(t)} + \left[-l''(\beta^{(t)})\right]^{-1} l'(\beta^{(t)}),$$

where $l'(\beta) = (\partial l/\partial \beta_1, \ldots, \partial l/\partial \beta_p)^T$ is the vector of first derivatives (also known as the score vector) and $l''(\beta)$ is the $p \times p$ matrix whose (j, k)th element is $\partial^2 l/\partial\beta_j\partial\beta_k$. The matrix $-l''(\beta)$ is also called the observed information. In some problems, the Newton-Raphson procedure becomes computationally simpler if the observed information is replaced by its expected value; the resulting algorithm is called Fisher scoring. In logistic regression, however, the expected and observed information are equal, so the Newton-Raphson and Fisher scoring methods are equivalent. Upon convergence, the inverse of this matrix evaluated at the mode $\hat\beta$ provides an estimated covariance matrix for the parameters,

$$\hat{V}(\hat\beta) = \left[-l''(\hat\beta)\right]^{-1},$$

and inferences are based on the approximation $(\hat\beta - \beta) \sim N(0, \hat{V}(\hat\beta))$. For more information, see Hosmer and Lemeshow (2000) or Agresti (2002).

Straightforward differentiation of (5.2) shows that

$$\frac{\partial l}{\partial\beta_j} = \sum_{i=1}^{N} x_{ij}(y_i - \mu_i),$$

$$\frac{\partial^2 l}{\partial\beta_j\partial\beta_k} = -\sum_{i=1}^{N} w_i x_{ij} x_{ik},$$

where $\mu_i = n_i\pi_i$ and $w_i = n_i\pi_i(1 - \pi_i)$. These can be succinctly written in matrix notation as $l'(\beta) = X^T(y - \mu)$ and $-l''(\beta) = X^T W X$, where $y = (y_1,\ldots,y_N)^T$, $\mu = (\mu_1,\ldots,\mu_N)^T$,

$$X = \begin{bmatrix} x_1^T \\ x_2^T \\ \vdots \\ x_N^T \end{bmatrix} = \begin{bmatrix} x_{11} & x_{12} & \cdots & x_{1p} \\ x_{21} & x_{22} & \cdots & x_{2p} \\ \vdots & \vdots & \ddots & \vdots \\ x_{N1} & x_{N2} & \cdots & x_{Np} \end{bmatrix},$$

and $W = \text{Diag}(n_i\pi_i(1 - \pi_i))$. The iteration of Newton-Raphson is

$$\beta^{(t+1)} = \beta^{(t)} + \left(X^T W^{(t)} X\right)^{-1} X^T \left(y - \mu^{(t)}\right), \tag{5.3}$$

where $W^{(t)}$ and $\mu^{(t)}$ are calculated by setting $\beta = \beta^{(t)}$. If we take $z_i = x_i^T\beta + (y_i - \mu_i)/w_i$ and $z = (z_1,\ldots,z_N)^T$, then (5.3) can be rewritten as

$$\beta^{(t+1)} = \left(X^T W^{(t)} X\right)^{-1} X^T W^{(t)} z^{(t)}, \tag{5.4}$$

showing that Newton-Raphson for the logistic model is equivalent to an iterative application of weighted least squares (WLS). In the literature of generalized linear models, z_i is often called the adjusted dependent variate or the working variate (McCullagh and Nelder, 1989).

To implement this procedure, we need a starting value and a stopping rule. It's usually good enough to begin at $\beta^{(0)} = 0$, so that the initial weights become $w_i \propto n_i$. We can stop iterating when the relative change in the elements of β becomes small; namely when

$$\left| \beta_j^{(t)} - \beta_j^{(t-1)} \right| \leq \epsilon \left| \beta_j^{(t-1)} \right| \tag{5.5}$$

holds for all $j = 1, \ldots, p$, where ϵ is a small positive number. Convergence is not guaranteed because the maximum-likelihood estimates for one or more elements of β may be infinite. To cover that possibility, we will declare a maximum number of iterations t_{max} and stop iterating when (5.5) occurs or when $t = t_{max}$.

5.5.3   Parameters and Results

Before implementing the model-fitting procedure, let's expand the ELOGIT session to hold estimates of the model parameters. Here is a derived type for that purpose.

```
 elogit_types.f90
!###################################################################
type :: param_type
   sequence
   private
   logical :: is_null = .true.
   integer(kind=our_int) :: p=0
   real(kind=our_dble), pointer :: beta(:)=>null()
end type param_type
!###################################################################
```

Because we are starting the algorithm at $\beta^{(0)} = 0$, we will not need to input any starting values. We will, however, need access to the results. Therefore, we have written a public method called `get_elogit_beta` that obtains the current value of β. A second method, `get_elogit_beta_names`, obtains character-string names for the elements of β that correspond to the names of the predictor variables. These procedures follow a now familiar pattern, so we do not show them here.

In addition to returning the final parameter estimate $\hat{\beta}$, we want to report how many iterations were carried out and whether the algorithm converged. We also want to report a few statistics to help the user draw inferences and assess the quality of fit. Standard errors for $\hat{\beta}$ come from the diagonal elements of $\hat{V} = (X^T W X)^{-1}$, so our model-fitting procedure should return this matrix. We should also return the value of the loglikelihood function (5.2) and measures that summarize the discrepancies between the observed responses y_i and their estimated means,

$$n_i \hat{\pi}_i = \frac{n_i \, e^{x_i^T \hat{\beta}}}{1 + e^{x_i^T \hat{\beta}}}.$$

The Pearson goodness-of-fit statistic is $X^2 = \sum_{i=1}^{N} r_i^2$, where

$$r_i = \frac{y_i - n_i\hat{\pi}}{\sqrt{n_i\hat{\pi}_i(1 - \hat{\pi}_i)}} . \tag{5.6}$$

Another common measure is the deviance statistic $G^2 = \sum_{i=1}^{N} d_i^2$, where

$$d_i^2 = 2\left\{ y_i \log\left(\frac{y_i}{n_i\hat{\pi}_i}\right) + (n_i - y_i) \log\left(\frac{n_i - y_i}{n_i - n_i\hat{\pi}_i}\right) \right\}. \tag{5.7}$$

As written, (5.7) is undefined when $y_i = 0$ or $y_i = n_i$; in those cases, one should interpret $0 \log 0$ as zero. If the n_i's are sufficiently large, then both X^2 and G^2 are approximately distributed as chi-square with $N - p$ degrees of freedom if the model holds; unusually large values of X^2 or G^2 relative to χ^2_{N-p} suggest model failure. For more discussion on the behavior of these fit statistics, consult McCullagh and Nelder (1989) or Agresti (2002). Here is our derived type for holding these results; public methods for getting the various properties are not shown.

```
───────────────── elogit_types.f90 ─────────────────
!###################################################################
type :: results_type
   sequence
   private
   logical :: is_null=.true.
   integer(kind=our_int) :: iter=0
   logical :: converged=.false.
   real(kind=our_dble), pointer :: cov_beta(:,:)=>null()
   real(kind=our_dble) :: loglik=0.D0, X2=0.D0, G2=0.D0
   integer(kind=our_int) :: df=0
end type results_type
!###################################################################
```

5.5.4 The Model-Fitting Procedure

Here is our procedure for fitting the logistic model. This procedure relies on repeated calls to `fit_wls`, the function for weighted least-squares regression developed in Section 4.4. By taking advantage of the new Fortran array operations and intrinsics, we were able to reduce the computational parts to just a few lines; most of the code is concerned with creating and destroying array workspaces and reporting errors.

```
───────────────── elogit_modelfit.f90 ─────────────────
!###################################################################
integer(our_int) function run_elogit_modelfit( session, err, &
      warn, maxits, eps) result(answer)
   ! Fits the logistic regression model by iteratively reweighted
   ! least squares, storing the final parameter value in
   ! param and the results in results.
```

```fortran
! If it fails for any reason, param and results are nullified.
implicit none
! declare arguments
type(elogit_session_type), intent(inout) :: session
type(error_type), intent(inout) :: err
type(error_type), intent(inout), optional :: warn
integer(kind=our_int), intent(in), optional :: maxits
real(kind=our_dble), intent(in), optional :: eps
! declare local variables and parameters
character(len=*), parameter :: subname = "run_elogit_modelfit"
real(kind=our_dble), allocatable :: y(:), n(:), x(:,:), w(:), &
    pi(:), z(:), beta(:), beta_old(:), log_odds(:), &
    odds(:), cov_beta(:,:), dev(:)
integer(kind=our_int) :: ncase, p, ijunk, status, j, posn
integer(kind=our_int) :: maxits0
real(kind=our_dble) :: eps0
character(len=12) :: sInt
real(kind=our_dble), parameter :: log_huge = &
    log( huge(real(0, kind=our_dble)) )
real(kind=our_dble) :: scale
!   check arguments
answer = RETURN_FAIL
if( session%dataset%is_null ) goto 700
if( session%model%is_null ) goto 710
! allocate local arrays
ncase = session%dataset%ncase
p = session%model%npred
if( session%model%intercept_present ) p = p + 1
if( p == 0 ) goto 720
allocate( y(ncase), n(ncase), x(ncase,p), w(ncase), pi(ncase), &
    z(ncase), beta(p), beta_old(p), log_odds(ncase), &
    odds(ncase), cov_beta(p,p), dev(ncase), stat=status)
if( status /= 0 ) goto 780
! fill in y, n, x
y(:) = session%dataset%data_matrix(:, session%model%y_col )
if( session%model%grouped ) then
   n(:) = session%dataset%data_matrix(:, session%model%n_col )
else
   n(:) = 1.D0
end if
posn = 0
if( session%model%intercept_present ) then
   posn = posn + 1
   x(:,posn) = 1.D0
end if
do j = 1, session%model%npred
   posn = posn + 1
   x(:,posn) = session%dataset%data_matrix( : , &
       session%model%pred_col(j) )
end do
! set maximum number of iterations and convergence criterion
if( present(maxits) ) then
   if( maxits < 0 ) goto 750
```

```
      maxits0 = maxits
else
      maxits0 = 20
end if
if( present(eps) ) then
      if( eps <= 0.D0 ) goto 760
      eps0 = eps
else
      eps0 = 1D-08
end if
! set starting value for beta
beta(:) = 0.D0
! main iteration
session%results%iter = 0
session%results%converged = .false.
do
      session%results%iter = session%results%iter + 1
      write(sInt,"(I12)") session%results%iter
      sInt = adjustl(sInt)    ! for error reporting
      beta_old = beta
      log_odds = matmul(x, beta)
      if( any( log_odds > log_huge ) ) goto 820   ! to prevent overflow
      odds = exp(log_odds)
      pi = odds / (1.D0 + odds)
      w = n * pi * ( 1.D0 - pi )
      z = log_odds + ( y - n*pi ) / w
      if( fit_wls( x, z, w, beta, cov_beta, scale, err ) &
            == RETURN_FAIL ) goto 840
      if( all( abs( beta - beta_old ) <= eps0*abs(beta_old) )) &
            session%results%converged = .true.
      if( session%results%converged .or. &
            (session%results%iter >= maxits0) ) exit
end do
! store the resulting estimate in param
session%param%is_null = .false.
session%param%p = p
if( dyn_alloc( session%param%beta, session%param%p, err) &
      == RETURN_FAIL ) goto 800
session%param%beta = beta
! store the covariance matrix in results
if( dyn_alloc( session%results%cov_beta, p, p, err) &
      == RETURN_FAIL ) goto 800
session%results%cov_beta(:,:) = cov_beta(:,:)
! calculate the fit statistics
session%results%loglik = sum( y*log(pi) + (n-y)*log(1.D0-pi) )
session%results%X2 = sum( (y-n*pi)**2 / w )
where( (y /= 0.D0) .and. (y /= n) ) &
      dev = y * log( y / (n*pi) ) &
            + (n-y) * log( (n-y) / (n-n*pi) )
where( y == 0.D0 ) dev = n * log( n / (n-n*pi) )
where( y == n ) dev = n * log( n / (n*pi) )
session%results%G2 = 2.D0 * sum(dev)
session%results%df = ncase - p
```

```
        session%results%is_null = .false.
        ! issue warnings, if warranted
        if( present(warn) ) then
           if( .not.session%results%converged ) &
                call err_handle( warn, 1000, &
                called_from = subname//" in MOD "//modname, &
                custom_1 = &
                "Did not converge by " // trim(sInt) // " iterations.")
        end if
        ! normal exit
        answer = RETURN_SUCCESS
        goto 999
        ! error traps
700     call err_handle(err, 1000, &
                called_from = subname//" in MOD "//modname, &
                custom_1 = "No dataset has been loaded yet.")
        goto 999
710     call err_handle(err, 1000, &
                called_from = subname//" in MOD "//modname, &
                custom_1 = "No model has been specified yet.")
        goto 999
720     call err_handle(err, 1000, &
                called_from = subname//" in MOD "//modname, &
                custom_1 = "Model has no parameters.")
        goto 999
750     call err_handle(err, 1000, &
                called_from = subname//" in MOD "//modname, &
                custom_1 = "Maximum no. of iterations cannot be negative.")
        goto 999
760     call err_handle(err, 1000, &
                called_from = subname//" in MOD "//modname, &
                custom_1 = "Convergence criterion must be positive.")
        goto 999
780     call err_handle(err, 200, &
                called_from = subname//" in MOD "//modname)
        goto 999
800     call err_handle(err, 1000, &
                called_from = subname//" in MOD "//modname)
        goto 999
820     call err_handle(err, 104, &
                called_from = subname//" in MOD "//modname, &
                custom_1 = "during iteration " // trim(sInt), &
                custom_2 = "Model fit procedure aborted.")
        goto 999
830     call err_handle(err, 102, &
                called_from = subname//" in MOD "//modname, &
                custom_1 = "during iteration " // trim(sInt), &
                custom_2 = "Model fit procedure aborted.")
        goto 999
840     call err_handle(err, 1000, &
                called_from = subname//" in MOD "//modname, &
                custom_1 = "during iteration " // trim(sInt), &
                custom_2 = "Model fit procedure aborted.")
```

```
        goto 999
        ! final cleanup
999     continue
        if( allocated(y) ) deallocate( y )
        if( allocated( n ) ) deallocate( n )
        if( allocated( x ) ) deallocate( x )
        if( allocated( w ) ) deallocate( w )
        if( allocated( pi ) ) deallocate( pi )
        if( allocated( z ) ) deallocate( z )
        if( allocated( beta ) ) deallocate( beta )
        if( allocated( beta_old ) ) deallocate( beta_old )
        if( allocated( log_odds ) ) deallocate( log_odds )
        if( allocated( odds ) ) deallocate( odds )
        if( allocated( cov_beta ) ) deallocate( cov_beta )
        if( allocated( dev ) ) deallocate( dev )
        if( answer == RETURN_FAIL ) then
           ijunk = nullify_elogit_session( session, err, &
                save_dataset = .true., save_model = .true.)
        end if
    end function run_elogit_modelfit
    !###############################################################
```

5.5.5 Reporting the Results

Our last procedure in this chapter takes the results from the model fit and writes a nicely formatted summary to an output file. Here is an example of the kind of output it produces.

```
                        _____ viral.out _____
    #########################################################
    #    ELOGIT                                             #
    #    A simple program for logistic regression analysis  #
    #    Version 1.0 - June, 2004                           #
    #    Written by J.L. Schafer                            #
    #    Department of Statistics and The Methodology Center #
    #    The Pennsylvania State University                  #
    #########################################################


    08 June, 2004
    14:52:04

    Data set information
       Number of cases:     5
       Number of variables:  3

       Variables
       -------------

       1  LOG_DOSE
       2  N
       3  Y

    Model specification
```

```
    Response (y):    Y
    Denominator (n): N
    Predictors:      INTRCPT
                     LOG_DOSE

Iteratively reweighted least-squares algorithm
  Converged at iteration 8

                 estimate      std.err.        ratio
              ------------   ------------   ------------

    INTRCPT     9.5868        3.7067         2.5864
    LOG_DOSE    2.8792        1.1023         2.6121

Summary of model fit

    Loglikelihood:    -6.7898028
    Deviance G^2:      0.53470113
    Pearson's X^2:     0.36098344

    Degrees of freedom: 3
```

The procedure for creating output files, which is shown below, calls two other procedures, which are not shown: `write_program_info_to_outfile`, which stamps the file with basic information about the ELOGIT program, and `write_date_and_time_to_outfile`, which reports the current date and time.

```
──────────────── elogit_io.f90 ────────────────
!######################################################################
integer(kind=our_int) function write_elogit_results_to_outfile( &
    output_file_name, session, err) result(answer)
  ! writes a formatted summary of results to the output file
  implicit none
  ! declare arguments
  character(len=file_name_length), intent(in) :: output_file_name
  type(elogit_session_type), intent(in) :: session
  type(error_type), intent(inout) :: err
  ! declare local variables and parameters
  integer(kind=our_int) :: i, ncase, nvar, iter, df
  logical :: converged
  character(len=var_name_length), pointer :: var_names(:)=>null()
  real(kind=our_dble), pointer :: beta(:)=>null(), &
      cov_beta(:,:)=>null()
  real(kind=our_dble) :: est, SE, ratio, loglik, X2, G2
  character(len=12) :: sInt, sRealA, sRealB, sRealC
  character(len=var_name_length), parameter :: var_blank=""
  character(len=*), parameter :: subname = &
      "write_elogit_results_to_outfile"
  ! check arguments
  answer = RETURN_FAIL
  if(output_file_name == "") goto 700
  ! open output file
```

```
open(unit=out_file_handle, file=output_file_name, &
    status="REPLACE", action="WRITE", err=800)
! stamp output file with program information
if( write_program_info_to_outfile(out_file_handle, err) &
    == RETURN_FAIL) goto 990
write(unit=out_file_handle, fmt="(A)", err=820) "" ! blank line
! stamp output file with time and date
if( write_date_and_time_to_outfile(out_file_handle, err) &
    == RETURN_FAIL) goto 990
write(unit=out_file_handle, fmt="(A)", err=820) "" ! blank line
! Write information about data set
if( get_elogit_ncase(ncase, session, err) == RETURN_FAIL) &
    goto 990
if( get_elogit_nvar(nvar, session, err) == RETURN_FAIL) &
    goto 990
if( get_elogit_var_names(var_names, session, err) == &
    RETURN_FAIL) goto 990
write(unit=out_file_handle, fmt="(A)", err=820) &
    "Data set information"
write(sInt,"(I12)") ncase
sInt = adjustl(sInt)
write(unit=out_file_handle, fmt="(A)", err=820) &
    "    Number of cases:      " // trim(sInt)
write(sInt,"(I12)") nvar
sInt = adjustl(sInt)
write(unit=out_file_handle, fmt="(A)", err=820) &
    "    Number of variables:  " // trim(sInt)
write(unit=out_file_handle, fmt="(A)", err=820) "" ! blank line
! List the variables
write(unit=out_file_handle, fmt="(A)", err=820) &
    "    Variables"
write(unit=out_file_handle, fmt="(A)", err=820) &
    "    -------------"
do i = 1, nvar
   write(unit=out_file_handle, fmt="(3X,I3,2X,A)", err=820) &
        i, trim( var_names(i) )
end do
write(unit=out_file_handle, fmt="(A)", err=820) "" ! blank line
! Report model specification
write(unit=out_file_handle, fmt="(A)", err=820) &
    "Model specification"
if( get_elogit_response(var_names, session, err) &
    == RETURN_FAIL ) goto 990
write(unit=out_file_handle, fmt="(A)", err=820) &
    "    Response (y):    " // trim(var_names(1))
if( size(var_names) > 1 ) then
   write(unit=out_file_handle, fmt="(A)", err=820) &
        "    Denominator (n): " // trim(var_names(2))
end if
if( get_elogit_beta_names(var_names, session, err) &
    == RETURN_FAIL) goto 990
if( associated(var_names) ) then
   write(unit=out_file_handle, fmt="(A)", err=820) &
```

```
              "   Predictors:        " // trim(var_names(1))
         do i = 2, size(var_names)
            write(unit=out_file_handle, fmt="(A)", err=820) &
                "                    " // trim(var_names(i))
         end do
     end if
     write(unit=out_file_handle, fmt="(A)", err=820) "" ! blank line
     ! report iteration details
     if( get_elogit_iter(iter, session, err) == RETURN_FAIL) goto 990
     if( get_elogit_converged(converged, session, err) &
        == RETURN_FAIL) goto 990
     write(unit=out_file_handle, fmt="(A)", err=820) &
         "Iteratively reweighted least-squares algorithm"
     write(sInt,"(I12)") iter
     sInt = adjustl(sInt)
     if(converged) then
         write(unit=out_file_handle, fmt="(A)", err=820) &
            "   Converged at iteration " // trim(sInt)
     else
         write(unit=out_file_handle, fmt="(A)", err=820) &
            "   Failed to converge by iteration " // trim(sInt)
     end if
     write(unit=out_file_handle, fmt="(A)", err=820) "" ! blank line
     ! report coefficients and standard errors
     if( get_elogit_beta(beta, session, err) &
         == RETURN_FAIL) goto 990
     if( get_elogit_cov_beta(cov_beta, session, err) &
         == RETURN_FAIL) goto 990
     write(unit=out_file_handle, fmt="(A)", err=820) &
         "   " // var_blank // "  " // &
         "  estimate      std.err.      ratio"
     write(unit=out_file_handle, fmt="(A)", err=820) &
         "   " // var_blank // "  " // &
         "------------  ------------  ------------"
     do i = 1, size(beta)
         est = beta(i)
         SE = sqrt( cov_beta(i,i) )
         ratio = est/SE
         write(sRealA,"(G12.5)") est
         write(sRealB,"(G12.5)") SE
         write(sRealC,"(G12.5)") ratio
         write(unit=out_file_handle, fmt="(A)", err=820) &
            "   " // var_names(i) // "  " // &
            sRealA // "  " // sRealB // "  " // sRealC
     end do
     write(unit=out_file_handle, fmt="(A)", err=820) "" ! blank line
     ! report goodness-of-fit measures
     if( get_elogit_loglik(loglik, session, err) &
         == RETURN_FAIL) goto 990
     if( get_elogit_X2(X2, session, err) == RETURN_FAIL) goto 990
     if( get_elogit_G2(G2, session, err) == RETURN_FAIL) goto 990
     if( get_elogit_df(df, session, err) == RETURN_FAIL) goto 990
     write(unit=out_file_handle, fmt="(A)", err=820) &
```

```
              "Summary of model fit"
       write(unit=out_file_handle, fmt="(A)", err=820) "" ! blank line
       write(unit=out_file_handle, fmt="(A20,G15.8)", err=820) &
              "    Loglikelihood:  ", loglik
       write(unit=out_file_handle, fmt="(A20,G15.8)", err=820) &
              "    Deviance G^2:   ", G2
       write(unit=out_file_handle, fmt="(A20,G15.8)", err=820) &
              "    Pearson's X^2:  ", X2
       write(unit=out_file_handle, fmt="(A)", err=820) "" ! blank line
       write(sInt,"(I12)") df
       sInt = adjustl(sInt)
       write(unit=out_file_handle, fmt="(A)", err=820) &
              "    Degrees of freedom: " // trim(sInt)
       ! normal exit
       close(unit=out_file_handle)
       answer = RETURN_SUCCESS
       return
       ! error traps
700    call err_handle(err, 1000, &
              called_from = subname//" in MOD "//modname, &
              custom_1 = "No output file name specified.")
       return
800    call err_handle(err, 2, &
              called_from = subname//" in MOD "//modname, &
              file_name = output_file_name)
       return
820    call err_handle(err, 4, &
              called_from = subname//" in MOD "//modname, &
              file_name = output_file_name)
       close(unit=out_file_handle)
       return
990    call err_handle(err, 1000, &
              called_from = subname//" in MOD "//modname)
       close(unit=out_file_handle)
       return
     end function write_elogit_results_to_outfile
     !###############################################################
```

In FORTRAN 77, the appearance of printed data was determined by format statements at compilation time. Print formats can now be determined at run time through a fmt string in a write statement. The old format statements still work, but use of the fmt string is preferred.

Style tip————————————————————————————————

Specify printing formats through fmt strings in your write statements.

The meaning of format strings has not changed since 1977. For example, X still denotes a blank space, A10 denotes a field of alphanumeric (character) data ten characters wide, I4 denotes an integer field four characters wide, and F12.5 indicates a floating-point number in a field of 12 characters, with five digits after the decimal. An integer before any of these specifiers indicates repetition: 5X denotes five blank spaces, 3I4 denotes three integers of four characters each, and so on. The decimal format F may cause a crash if the number to be printed is too large or too small. In lieu of F, we prefer to use the new specifier G, which automatically switches between decimal and exponential notation depending on the magnitude of the argument.

After completing the output file routines, two more details are needed to get ELOGIT working. First, we need to add sections to the control file that allow the user to specify the maximum number of iterations t_{max}, the convergence criterion ϵ, and the name of the output file. Then we need to call run_elogit_modelfit and write_elogit_results_to_outfile from the main program. These additions have been incorporated into the final version of ELOGIT, which is available from our Web site.

5.5.6 Looking Ahead

Throughout this chapter, we have emphasized that statistical programming is more than producing good computational routines. A program should be robust, able to handle all kinds of errors without crashing. It should be intelligent, providing informative feedback to enable users to correct their own mistakes. It should also be extensible, allowing the programmer to add new features incrementally without rewriting existing code.

The source for the ELOGIT console program resides in 14 files and occupies approximately 3,700 lines. Only 10% of these lines, however, have anything to do with printing to the screen or reading commands from the keyboard and control file. Console-related features are confined to the elogit_ctrlfile module and to the main program in elogit.f90. If we simply remove these two files, the elogit_types module remains intact as an intelligent, self-contained engine for logistic regression that could easily plug into other Fortran console programs. It could also interact with a graphical interface or with many programs already installed on your computer—provided that all the tricky details of interoperability can be worked out. We will show how to do this in the chapters ahead.

5.6 Additional Exercises

1. Add get methods to the elogit_types module to obtain Pearson and deviance residuals. The Pearson residuals are given by (5.6), and the deviance residuals are

$$d_i = \text{sign}(y_i - n_i\hat{\pi}_i)\sqrt{d_i^2},$$

where d_i^2 is defined in (5.7).

2. The Pearson goodness-of-fit statistic X^2 and the deviance G^2 are often used to compare the fit of nested logistic models. Suppose that model A is a special case of model B (e.g., the predictor variables in A are a subset of those in B). If X_A^2 and X_B^2 denote the Pearson statistics for the two models, and if the smaller model A is true, then the difference

$$\Delta X^2 = X_A^2 - X_B^2$$

is approximately distributed as chi-square with degrees of freedom equal to the number of free parameters in B minus the number of free parameters in A. The distribution of the difference in deviance statistics,

$$\Delta G^2 = G_A^2 - G_B^2,$$

approaches the same distribution and is asymptotically equivalent to ΔX^2. Modify ELOGIT to compute and print out the values of ΔX^2 and ΔG^2 for comparing the fit of the user-specified model to that of an intercept-only model with no predictors.

3. An important feature of the binomial model is that the mean $E(y_i) = \mu_i = n_i \pi_i$ and the variance $V(y_i) = n_i \pi_i (1 - \pi_i)$ are related by

$$V(y_i) = \mu_i(n_i - \mu_i)/n_i. \tag{5.8}$$

Because real data may violate this relationship, the model is often expanded to

$$V(y_i) = \sigma^2 \mu_i(n_i - \mu_i)/n_i,$$

where $\sigma^2 > 0$ is a scale parameter (McCullagh and Nelder, 1989). Introduction of a scale parameter does not affect the estimation of β, but if σ^2 is substantially different from 1.0, the estimated covariance matrix for $\hat{\beta}$ should be replaced by $\sigma^2(X^TWX)^{-1}$, and the goodness-of-fit statistics X^2 and G^2 should be replaced by X^2/σ^2 and G^2/σ^2. The traditional way to estimate the scale parameter is by fitting a "maximal model" (i.e., a model that contains all the covariates worth considering) and taking $\hat{\sigma}^2 = X^2/(N - p)$ under this model. Modify ELOGIT to allow three different options for the scale parameter: set σ^2 equal to 1.0 (the default); set σ^2 equal to a user-supplied value; or estimate σ^2 from the current model, treating it as the maximal model.

4. An alternative estimated covariance matrix for $\hat{\beta}$ in the logistic model, using the notation of (5.3), is

$$\tilde{V}(\hat{\beta}) = (X^TWX)^{-1}(X^TWAWX)(X^TWX)^{-1},$$

where $A = \mathrm{Diag}((y_i - \mu_i)^2)$. This estimator, called the information sandwich, retains its consistency when the variance relationship (5.8) is violated (Huber, 1967; White, 1980). Modify ELOGIT to compute and print standard errors based on $\tilde{V}(\hat{\beta})$ upon request.

5. An EM algorithm for estimating the parameters from a mixture of two exponential distributions was presented in Section 4.2. Turn this computational procedure into an object-oriented Fortran module, and use the module from a console application.

6. Write a complete console application for performing the chi-square test for independence in a two-way contingency table following the object-oriented strategies described in this chapter.

6
Creating and Using Dynamic-Link Libraries

In Chapters 2–5, we have shown how to write procedures and programs in pure Fortran 95. The focus of this book now shifts to interoperability—how to create Fortran procedures that interact with other applications on your computer.

In this chapter, we describe how to turn Fortran procedures into dynamic-link libraries (DLLs) in Windows or shared objects in Unix or Linux. DLLs and shared objects provide a convenient way to add powerful and fast computational routines to statistical packages such as SAS, S-PLUS, and R. Unfortunately, each package calls and uses them in a different way. If your goal is to quickly implement a computational procedure for one of these environments, then a DLL may be a sensible approach. If you want to create one set of procedures that can be called by multiple applications, it may be better to create a COM server, as we will discuss in Chapter 7. In the latter case, we still recommend that you peruse this chapter, because it will introduce you to many key issues of interoperability.

6.1 Extending the Functionality of Statistical Packages with Fortran DLLs

6.1.1 Compiled Procedures Run Faster

Statistical packages such as SAS, S-PLUS, and R maintain large libraries of built-in procedures for regression and other popular data analysis techniques. For nonstandard analyses, however, users must write their own code. In SAS, this is usually done through a combination of data steps, the SAS macro language, and PROC IML. Users of S-PLUS and R may write their own functions in the S language. Nevertheless, the developers of these environments have also recognized that users may occasionally want to call computational procedures written in C or Fortran.

The main benefit of calling external procedures is speed. A routine that has been compiled directly from C or Fortran may run much faster—in some cases, orders of magnitude faster—than the same routine written in SAS, S-PLUS, or R. The reason is that the latter languages are interpreted; that is, the programmer's code is converted into low-level instructions understood by the computer's processor while the program is running. This interpretation can be especially onerous for nested loops because the same code is reinterpreted at each cycle. If the statistical package can invoke an external procedure that has already been compiled, the same computations can be done much more efficiently.

6.1.2   When to Use a DLL

One way to make a Fortran procedure available to one of these programs is to package it as a conventional dynamic-link library (DLL). DLLs are a fundamental tool of modern software development, the building blocks of applications that run in Windows and of the Windows operating system itself. In Unix and Linux systems, these components are called shared objects. With DLLs, multiple programs can share a common set of procedures without any duplication of the compiled code.

In the remainder of this chapter, we show how procedures written in Fortran can be encased in a conventional DLL and called from popular statistical packages. Many of these packages—which themselves are built upon DLLs—provide a way for users to interact with additional DLLs. Unfortunately, the devil is in the details; subtle differences in the way data are handled by these programs make it difficult or impossible for one Fortran DLL to serve them all. We show by example how to create DLLs and invoke them from SAS, S-PLUS, and R. We focus on these three environments because of their popularity among statisticians and biostatisticians. Recent versions of Stata and MATLAB can also call DLLs; consult those programs' documentation for details.

Because conventional DLLs are not object-oriented, the kinds of procedures that can be called through them are somewhat limited. DLLs are effective for manipulating data passed as arguments in a single procedure call, but less effective for storing data that must persist from one call to the next. Arrays of variable sizes may need to be passed in the FORTRAN 77 style, with the dimensions also passed as arguments. Array arguments may not be redimensioned within a procedure, so the size of each output must be known in advance. Derived types may be used as local variables within the procedure but may not appear in the argument list . In some cases, you may encounter difficulties in passing character strings, and the total number of arguments may be restricted.

When using DLLs, programming mistakes can be difficult to diagnose and fix. A conventional DLL does not spawn a new process on your computer; rather, it loads new procedures into the program that is already running. If any problem arises—e.g. because the procedure is not called properly or arguments are passed incorrectly, or because of a bug in the procedure itself—then the whole application may crash, with little information given as to why. Getting a DLL to work with a statistical package will inevitably involve some trial and error.

Because of these limitations, we recommend conventional DLLs when you want to quickly develop an external computational procedure for a single statistical package. Conventional DLLs are a good choice for calling the kind of non-object-oriented computational procedures discussed in Chapter 4, whose arguments are scalars and arrays. They are less appropriate for object-oriented modules such as the `elogit_types` module developed in Chapter 5. If your module is object-oriented, it will be much easier to turn it into an in-process or out-of-process COM server, as we will discuss in the chapters ahead.

Before you start to design and develop a DLL, we strongly recommend that you thoroughly read the documentation on DLLs accompanying your compiler. This chapter will provide general principles and a few simple examples, but we cannot go into great detail on any specific compiler. It's a good idea to first create a very simple DLL for testing purposes to see which options are necessary to make a DLL work with your statistical package of interest. If you have not yet selected a compiler, you may wish to contact other programmers who have developed similar applications and get their recommendations about which ones work best.

6.2   Understanding Libraries

6.2.1   Source-Code Libraries

In the language of programming, a library is simply a collection of procedures. The procedures in a library are usually related in some way. For example, LAPACK is a well-known library of high-performance algorithms for linear algebra. LAPACK itself relies on another library called BLAS, a collection of low-level vector and matrix routines.

The most basic kind of Fortran library is a single source-code file, or a collection of source-code files, containing Fortran functions and subroutines. One who develops and distributes a source-code library assumes that the user is a programmer with access to a Fortran compiler. Distributing source code has certain advantages; the user can view the inner workings of the routines and customize and perhaps improve them. In many cases, however, authors prefer to distribute their libraries in a way that prevents the procedures from being dissected or modified, or in some form that does not require the end user to possess a compiler. That is, many developers prefer to distribute libraries that have already been compiled. There are different kinds of compiled libraries, as we now describe.

6.2.2   Static Libraries

One kind of compiled Fortran library is the static object file. Static object files are pieces of binary code that have yet to be assembled—or to use the correct term, linked—into an executable program. In Windows systems, these files are usually named with the .obj suffix; in other platforms they are called *.o. The process of linking object files into an executable program is called static linking. After static linking, the object files are no longer needed to run the program because all of the binary code within these files has been embedded within the executable image.

Each Fortran source-code file that is compiled produces one static object file. Sometimes object files are grouped together into collections called static libraries. Static library files, which in Windows systems have the .lib suffix, help developers to better manage their code. When you link to procedures within a static library, only the binary code for those procedures that are actually needed is embedded within the executable image, which makes the executable program smaller. If a procedure is shared by several programs, each executable file will contain its own copy of the binary code. Object-code files and libraries created by different compilers are usually incompatible and cannot be linked together. Even different versions of the same compiler may produce incompatible object-code. For these reasons, object code files and libraries are used mainly by individual programmers or by members of a development group who are using the same compiler.

In most cases, compiling Fortran routines into object code is not the best way to make them accessible to other programs on your computer. This is well-illustrated by earlier versions of S-PLUS. Prior to Version 6, S-PLUS for Windows was capable of linking to external Fortran subroutines contained in object code (.obj) files, provided that (a) they were written in FORTRAN 77, (b) had no file I/O or print statements, and (c) were compiled using the Watcom compiler, which was commercially available at that time. Watcom Fortran was discontinued in 1999, creating a dilemma for programmers who were using this approach. Since then, the developers of S-PLUS have abandoned that strategy in favor of DLLs.

6.2.3   Dynamic-Link Libraries

Unlike static libraries, DLLs are both compiled and linked to produce a special kind of library file. In Windows systems, these files have the .dll suffix; on Unix and Linux systems, the extension is .so. A DLL does not run on its own but must be called by other executable programs. As the name suggests, the linking to these programs is performed dynamically by the operating system while the program is running. That is, dynamic linking occurs when the user runs programs that need one or more of the library's procedures. The binary code within the DLL is loaded into memory only when it is needed—when a running program needs to call procedures in the DLL.

DLLs promote reuse of code at the binary level. One DLL on your hard disk may serve many different programs installed on the computer. Programs running at the same time may even share a single instance of the code in memory. Because the DLL code is not embedded in the executable file, the image of a complex program that relies on hundreds or thousands of library procedures may remain quite small and able to be loaded into memory and started up quickly.

The Windows operating system is built upon DLLs, and so are Windows applications. The graphical routines that display windows, menus, text boxes, push buttons, dialog boxes, directory trees, etc. all reside in various DLLs, so that programs developed for the Windows environment can share them and maintain a common look and feel. DLLs are also helpful for software maintenance and support; developers can fix bugs and provide enhancements by distributing new versions of specific DLLs instead of requiring users to reinstall the entire application. Because DLLs do not contain source code, they are not easily analyzed or modified, so developers of proprietary software are able to distribute them rather freely and yet maintain control over their intellectual property.

The free-wheeling nature of DLLs also causes trouble if they are not applied with care. A basic understanding of how DLLs work is essential to help you avoid common pitfalls.

6.3   How Programs Use DLLs

6.3.1   Locating the DLL

When a program wants to invoke a routine in a DLL, it must first locate the DLL file in which it resides. In order to be found, a DLL must sit in a directory where the program will look for it. This may include

- the directory in which the program's executable (.exe) file has been installed;

- the current working directory from which the program has been invoked (e.g. when the program is started from a command prompt);

- the directory in which the operating system has been installed; or

- any directory pointed to by the PATH environment variable.

If a program cannot locate a DLL that it needs, the operating system will display an error message at the command line or in a dialog box.

DLLs are identified solely by filename. It is quite possible that unrelated DLLs with different functionalities may be given the same name; in that case, a failure to locate the correct DLL may cause a program to crash.

6.3.2 DLL Hell

The practice of placing DLLs in the operating system's installation directory has led to many annoying problems. Suppose that a user installs a new program on a computer, and during the installation process a library named matrix.dll is copied into the Windows directory. Later, the user installs another application with a library named matrix.dll, and that file is copied into the Windows directory, overwriting the first one. The first application will no longer work properly. This problem can be avoided by placing DLLs used by a specific program into that program's own installation directory.

More subtle problems may arise when a DLL is shared by multiple programs. Suppose, for example, that two applications share a common set of numerical algorithms that have been packaged in a DLL. Moreover, suppose that one of the applications is distributed with a more recent version of the DLL bearing the same name as an older version. This new version may not be backwardly compatible with the other application and may cause it to crash. In that case, each application may need to have its own version of the DLL, placed in a directory where the application that needs it will find it first.

Problems such as these have become known as "DLL hell." Newer development paradigms based on COM solve this problem by automatically labeling libraries with elaborate identifiers guaranteed to be unique.

6.3.3 Dynamic Loading and Linking

A running computer program is a set of binary instructions. These instructions must be loaded into the computer's memory, and the processor executes them sequentially as they are arranged in memory. When the processor encounters a procedure call, the point of execution jumps to a new location in memory determined by its memory address.

If a large number of programs are running at the same time, these binary instructions and the data they operate upon may exceed the available physical memory on your RAM chips. In that case, the overflow is stored in temporary paging files—also known as swap space—on your computer's hard drive. The physical memory on the chips and the swap space on the hard drive are collectively known as the virtual memory space.

When a program is first invoked, the binary instructions from the executable (.exe) file are loaded into memory and execution begins. When the program calls a procedure in a DLL, binary code from the .dll file is loaded into the same compartment of virtual memory as the running program; this is called *dynamic loading*.

Not all of the procedures within a DLL may be invoked by a running program. Some procedures are private in the sense that they may only be called by other procedures in the DLL; others are publicly accessible to programs and other DLLs. The latter procedures are said to be *exported*. Once a DLL has been dynamically loaded, the running program may access an exported procedure of a DLL. To do so, however, it must know the memory address that marks the starting point of that procedure. The memory locations of the exported procedures are called *entry points*. The process of obtaining addresses of the entry points at run time is called *dynamic linking*.

6.3.4   Load-Time and Run-Time Linking

There are two types of dynamic linking. The first, which is called *load-time dynamic linking*, is used mainly by software developers. To perform load-time dynamic linking, you must have a complete, working, and compiled version of the DLL on hand before you compile the program that will use it. Along with the .dll file, you must also have an import library or .lib file. (Although this has the same filename extension as a static library file, its purpose is quite different.) Most compilers capable of producing DLLs will generate the .lib file automatically or upon request. The import library contains special code that will be embedded into the executable program, allowing the program to link with the DLL's exported procedures the moment it is loaded into memory at run time. Because users of pre-existing applications such as SAS and S-PLUS normally do not have access to the program's source code, load-time linking is usually not appropriate for extending those applications. (Developers of the R package have made their source code available; however, not many users compile and link their own installations of R.)

The second approach, called *run-time dynamic linking*, does not require an import library or access to the application's source code. This is the preferred way to call an external Fortran procedure from SAS, S-PLUS, or R. Run-time dynamic linking works in the following manner. While the main application is running, it issues a request to the operating system to load the DLL into memory. Next, it issues another command to locate the DLL's entry points. These are gleaned from the DLL's *export table*, a section of binary code in the .dll file that contains a list of all the exported procedures and the relative memory locations of their entry points. From these relative locations, the main program is able to deduce the actual locations of the entry points in the virtual memory space. When these actual locations are computed and stored by the main program, dynamic linking is complete.

Fortunately for us, the messy details of run-time linking and determination of the entry points are handled by the operating system. From our perspective, the issues that matter are: how to produce a DLL with the correct procedures listed in its export table; how to dynamically link the DLL with SAS, S-PLUS, or R; how to invoke an exported procedure from the statistical package; and how to make sure that data are passed correctly between the statistical package and the exported procedure.

6.4   Creating a Fortran DLL

6.4.1   The Basic Steps

Every major Fortran compiler for the Windows platform is capable of producing DLLs. The procedure for doing this varies somewhat depending on which compiler you are using and whether you compile at the command line or use an Integrated Development Environment (IDE). In every case, you must begin with one or more Fortran source-code files. Then you must identify the functions and subroutines to be exported. This may be done, for example, by including special comment lines read by that compiler. Then you compile and link the code using specific options that indicate the product should be a DLL rather than an ordinary object-code file or executable program.

This process sounds simple enough. When doing this, however, we need to pay special attention to how the exported procedures are designed, how the exports are named, and how arguments are passed.

6.4.2   Passing Arguments

A DLL written in Fortran may be used by an application written in another language. To ensure that arguments are passed correctly from one language to another, we need to understand how the languages store data. In particular, because most of the major statistical packages are written in C, we need to understand some differences between Fortran and C.

Character Strings

A character string in the C language is a sequence of one-byte alphanumeric characters, with the final character being a null or zero code. Fortran does not use the null terminator. This discrepancy may cause problems because character-string arguments are usually passed by reference rather than by value. Passing by reference means that, rather than sending an actual copy of the character string into a procedure, the computer passes only the address in memory where the string begins. When a Fortran procedure finishes running and returns control to a main program written in C, only the address of the string argument is passed back to the main program. As a result, the main program will locate the beginning of the string, but it will not necessarily know where the string ends; it will be expecting a null terminator that may not be there.

One way to solve this problem is to avoid character-string arguments altogether. There are many situations where passing strings is beneficial, however (e.g. when reporting error messages). Another strategy is to pass the length of the string to the DLL export as an additional integer argument. Another possibility is to fix the length of the string in advance to some constant value. The interface to Fortran provided by the R language assumes that all string arguments have a length of 255 bytes. The best way to handle character strings may vary from one platform to another; we will return to this issue shortly.

Integer and Floating-Point Variables

By now you are well-aware that Fortran compilers provide different kinds of integer and floating-point real variables. Windows operating systems have different types as well. In Windows, an integer value may be short (16 bits) or long (32 bits), and a floating-point value may have single precision (32 bits) or double precision (64 bits). When integer or real data are passed from a main program to a DLL and back again, discrepancies in assumed length can be disastrous. To help prevent problems, we recommend that you standardize your Fortran code to use 32-bit integers and 64-bit reals.

Arrays

Passing arrays between a main program and a DLL may also be problematic. Arrays, like character strings, usually need to be passed by reference; that is, the program and the DLL communicate to each other the memory address of where the array begins rather than the data values stored in the array. In most cases, it will also be necessary to pass the dimensions of the array as integer arguments.

6.4.3   Calling Conventions

When creating and using a DLL, you must also correctly specify the calling conventions. Failure to do this may result in, at best, a DLL that simply does not work and, at worst, one that crashes your computer.

One might wonder why DLL calls were never standardized to a single convention. Calling conventions originated in the linking of static objects. The authors of compilers devised a variety of rules by which routines were identified and invoked so that a linker could connect them at build time. Even though the compilers had different conventions, incompatibilities were irrelevant because each compiler had its own linker; a universal standard was unnecessary. With the advent of dynamic linking, these differences became an issue. Nevertheless, applications for the Windows platform typically adhere to a few widely accepted conventions, with only minor variations.

Passing Arguments by Reference or by Value

As we have already discussed, arguments are passed between a procedure and the calling program either by reference or by value. By reference means that a memory location is passed, and by value means that a copy of the contents is passed. A calling program expects the DLL to adhere to one method or the other, so it is important to make sure that the DLL conforms. S-PLUS and R always pass arguments by reference, whereas SAS can do it either way.

Symbolic Names

Symbolic names identify the procedures in the DLL's export table. Sometimes a procedure's symbolic name is the same as its name in the Fortran source. But some calling conventions require that an underscore (_) be appended to the end of the symbolic name. Symbolic names are also case-sensitive, whereas the Fortran language is not; hence there is ambiguity as to whether the characters in the symbolic name should be all lowercase, all uppercase, or exactly as they appear in the source code. Some compilers automatically append numerals to the symbolic name representing the number of bytes occupied by the procedure arguments.

Compilers that produce DLLs may provide a way for you to change the form of the symbolic names. They may also provide utilities that display information in a DLL's export table, so that you can see the symbolic names of the exported procedures in the DLL that you create. For example, Microsoft's Visual Studio IDE comes with a utility called dumpbin. If you have a DLL called matrix.dll, then typing

```
dumpbin /exports matrix.dll
```

at a command prompt will display a list of the exported symbols.

The symbolic names of your exports must match the names that the calling program is expecting. Details of what the program expects are not always well-documented and may require a bit of research. In S-PLUS and R, calls to Fortran DLLs are made through a function called .Fortran(). Each of those environments expect an underscore to be appended to each symbolic name. A lesser-known function, symbol.For(), will display the expected symbolic name corresponding to any Fortran procedure name. In R, the user is allowed to specify a custom naming convention.

Determining String Length

When a character-string argument is passed by reference to a DLL procedure, some Fortran compilers will automatically append the length of each character string to the string's address in virtual memory. Some applications that work with Fortran DLLs may use this mechanism, but others do not. You will need to consult your compiler's documentation to find out how to create a DLL that handles this issue properly.

6.4.4 Compiling and Linking the Source Code

When compiling your source code to create a DLL, you must indicate to the compiler which procedures are to be exported. The mechanism for doing this varies from one compiler to another.

With Intel Visual Fortran Version 8.0, an export is identified by a special compiler directive contained in a Fortran comment line. The directive looks like this.

```
subroutine mysub( arg1, arg2 )
   !DEC$ ATTRIBUTES DLLEXPORT :: mysub

   ! body of the procedure goes here

end subroutine mysub
```

A comment line beginning with `!DEC$ ATTRIBUTES` is a special directive that will be read by the Intel compiler but ignored by other compilers. In this case, the attribute `DLLEXPORT` is being applied to the subroutine `mysub`; as a result, the compiler will export this procedure using its default settings. By default, the symbolic name is the Fortran procedure name written in uppercase, as in `MYSUB`. This symbolic name can be changed by optional keywords when compiling at a command prompt; `/names:lowercase` makes it lowercase, and `/names:as_is` makes it case-identical to the procedure name in the Fortran source code. With additional keywords and directives, you can change the calling conventions and the manner in which any or all arguments are passed.

In Lahey/Fujitsu Fortran 7.0, exports are identified not by a comment line but by an actual statement in the Fortran code.

```
subroutine mysub( arg1, arg2 )
   dll_export MYSUB

   ! body of the procedure goes here

end subroutine mysub
```

The `dll_export` statement is recognized only by the Lahey/Fujitsu compiler. The argument `MYSUB`, which is case-sensitive, forms the basis of the symbolic name in the export table.

6.4.5 Compiler Options

When compiling and linking your code, you will need to notify the compiler that the intended product is not an object code (`.obj`) or executable (`.exe`) file but a `.dll` file. You may also be able to change the properties of the exported procedures while compiling and linking. If you are invoking Intel Fortran from a command prompt, the command-line option `/DLL` instructs the compiler to create a DLL with an export table based on the information provided by `!DEC$ ATTRIBUTES` compiler directives in the source code. The corresponding option in the Lahey/Fujitsu compiler is `-dll`. Lahey/Fujitsu also provides an option `-ml`, which stands for "multiple language," to make the calling conventions of the exports compatible with a specific language. The option `-ml msvc` produces a DLL that should work with any program compiled with Microsoft Visual C++. Intel Fortran has a similar option, `/iface`; the setting `/iface:cref` specifies that all exports will use the C-language by-reference conventions.

  If you are developing your DLL in the Microsoft Visual Studio IDE, you can access these options through the menu sequence Project → *project-name* Properties, which will open the project properties dialog box. We will demonstrate this by an example shortly.

6.5 Example: a Fortran DLL for Fitting an Exponential Mixture

In Section 4.2, we created a Fortran module for fitting a two-component mixture model. We now show how to package this module, and the modules used by it, as a DLL and call the model-fitting routine from SAS, S-PLUS, and R. The procedure for creating the DLL is compiler-dependent. We will demonstrate the process with Intel Visual Fortran Version 8, invoking the compiler both through the command line and through the Microsoft Visual Studio IDE. We will also briefly describe how to do it with Lahey/Fujitsu Fortran Version 7 and Salford FTN95 Version 4.5.

6.5.1 Creating a Wrapper

The procedure `run_em_exponential` was designed to be called from a Fortran program or procedure. It would not be appropriate, however, to export this procedure with a DLL and call it from a statistical package. One reason why we should not call `run_em_exponential` directly is that it is a public function contained within the module `em_exponential_engine`; any program calling the function would need to use that module, but the concept of using a Fortran module is not known to other languages. Another important reason why we should not call `run_em_exponential` is that one of its arguments is an `error_type`, a custom object defined in the `error_handler` module. Creating a compatible data structure in SAS, S-PLUS, or R to match this argument would be difficult at best.

For these reasons, we will create a *wrapper* for the `run_em_exponential` function and export the wrapper to the DLL. The wrapper is merely a procedure written in Fortran that "wraps around" the function, transferring data between the calling program and the function through its arguments. The wrapper will be an external procedure, not contained in any module, although it will use the `em_exponential_engine` and `error_handler` modules and create an instance of the `error_type` as a local variable. If an error occurs when the wrapper calls `run_em_exponential`, the error message will be written to a character string and passed to the calling program. All of the wrapper's arguments will be standard data types—long integers, double-precision reals, and character strings.

Because the wrapper needs to interact with a statistical program, it is important to consult that program's dcoumentation to review its requirements. According to the *S-PLUS 6 for Windows Programmer's Guide* (Insightful Corp., 2001), DLL procedures called from S-PLUS must be subroutines, not functions. Individual character strings may be passed as arguments, but arrays of strings are not allowed. Moreover, the Fortran procedure should not contain any `write` or `print` statements because the Fortran I/O unit numbers may conflict with unit numbers currently in use by S-PLUS. (In our experience, opening files and writing to them usually does work and is occasionally very useful.) The document *Writing R Extensions* (R Development Core Team, 2004) mentions similar restrictions and adds one more: all character-string arguments should be 255 bytes long (`len=255`). S-PLUS and R both expect arguments to be passed by reference. SAS, on the other hand, provides the user with more flexible calling and argument-passing capabilities; these are described in *SAS Companion for the Microsoft Windows Environment, Version 8* (SAS Institute, 2000). Using Intel Fortran Version 8 with Visual Studio, we are able to create one wrapper that will work with all three of these environments.

To begin, we start Visual Studio .NET and use the menus to create a new Fortran project. A New Project dialog window appears, as shown in Figure 6.1, and we select "Dynamic-Link Library" as the template. We name this project `em_exp`, so that the resulting DLL file will be `em_exp.dll`. (SAS requires the DLL file's base name (i.e., without the `.dll` extension) to be no more than eight characters long, so we have adhered to this constraint.) After selecting "OK," a second dialog box, known as the "Fortran DLL Project Wizard" appears (Figure 6.2). In that dialog, we select "Empty project" and click "Finish."

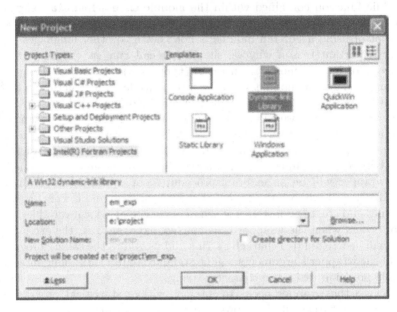

FIGURE 6.1. Intel Fortran New Project dialog.

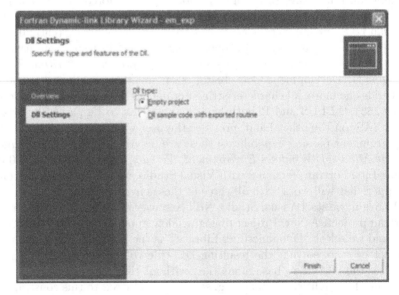

FIGURE 6.2. Intel Fortran DLL Project Wizard.

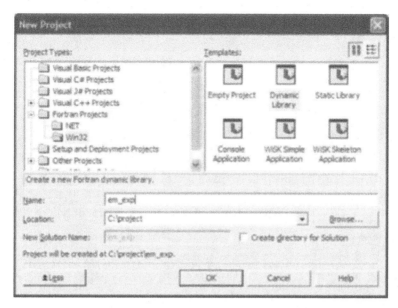

FIGURE 6.3. New DLL project dialog for Lahey/Fujitsu Fortran.

Recent versions of Lahey/Fujitsu and Salford Fortran also work with Visual Studio, and the procedure for those compilers is very similar. In Lahey/Fujitsu Fortran, we would create a "Dynamic Library" project under "Win32" (Figure 6.3); in Salford, we would select "FTN95 Application Extension" (Figure 6.4).

The existing code for this project resides in three modules, so we need to add the source code for these modules to our project. In the Solution Explorer, we simply add `constants.f90`, `error_handler.f90`, and `em_exponential_engine.f90` as "Source Files." Next, we add a new Fortran source-code file to the project, named `em_exponential.f90`, to hold the wrapper subroutine. The source code for our wrapper is shown below.

```
————————————— em_exponential.f90 —————————————
!#####################################################################
subroutine em_exponential(n, y, maxits, eps, pi, lambda_1, lambda_2, &
    iter, converged, loglik, score, hessian, msg_len, msg)
  !DEC$ ATTRIBUTES DLLEXPORT :: em_exponential
  use error_handler
  use program_constants
  use em_exponential_engine
  implicit none
  ! declare arguments
  integer, intent(in) :: n
  real(kind=our_dble), intent(in) :: y(n)
  integer, intent(in) :: maxits
  real(kind=our_dble), intent(in) :: eps
  real(kind=our_dble), intent(inout) :: pi, lambda_1, lambda_2
```

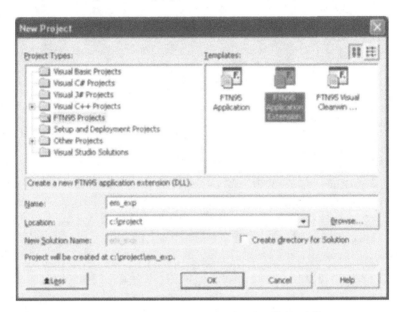

FIGURE 6.4. New DLL project dialog for Salford Fortran.

```
    integer, intent(out) :: iter
    logical, intent(out) :: converged
    real(kind=our_dble), intent(out) :: loglik, score(3), hessian(3,3)
    integer, intent(in) :: msg_len
    character(len=msg_len) :: msg
    ! declare locals
    integer(kind=our_int) :: ijunk
    type(error_type) :: err
    ! begin
    call err_reset(err)
    ijunk = run_em_exponential( y, pi, lambda_1, lambda_2, &
      iter, converged, loglik, score, hessian, err, maxits, eps )
    if( err_msg_present(err) ) call err_get_msgs(err, msg, "UNIX")
    return
end subroutine em_exponential
!#################################################################
```

As you can see, the wrapper does very little; it merely calls the model-fitting procedure and reports any messages. The most important aspect of this wrapper is its argument list. All integers are long (32 bits), and the reals are double precision (64 bits). Arrays of assumed length are not allowed in an exported DLL procedure, so the dimensions of each array are either fixed or passed as arguments. Character-string arguments of assumed length are not allowed either, so the length of the string msg is also passed as an argument. The intent attributes are allowed and are chosen appropriately. The only unusual feature is the compiler directive in the comment line,

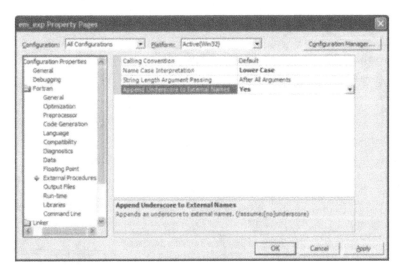

FIGURE 6.5. Intel Fortran Project Property settings.

which tells Intel Fortran to export this subroutine to the DLL. If we were using Lahey/Fujitsu, we would need to add the following line below the `implicit none` statement:

```
dll_export em_exponential
```

6.5.2 Building the DLL with Intel Visual Fortran and Lahey/Fujitsu Fortran

When building the DLL, we need to set a few important properties. First, we must give the exported procedure a proper symbolic name. S-PLUS and R expect an appended underscore, whereas SAS is more flexible. To conform to all three environments, let us choose `em_exponential_` as the symbolic name. To instruct the Intel compiler to use this convention, we use the menu to open the project's Property Pages dialog:

Project → em_exp Properties. . .

A dialog box will appear as shown in Figure 6.5. In the Configuration combo box at the upper-left corner, select "All Configurations." In the tree displayed on the left-hand side, select "External Procedure" under "Fortran"; this category defines the calling conventions. Change the "Name Case Interpretation" setting to "Lower Case" and the "Append Underscore to External Names" setting to "Yes." These settings have the same effect as the `/names:lowercase` and `/assume:underscore` options when compiling at a command prompt.

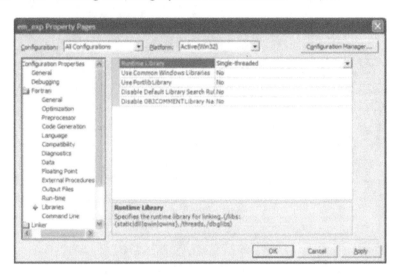

FIGURE 6.6. Selecting the single-threaded runtime library.

A few more Intel project properties must also be changed in order for the DLL to work. First, we need to change the default selection of a multithreaded runtime library to one that is single-threaded. To do this, select "Libraries" in the tree on the left-hand side, and change the "Runtime Library" setting to "Single-threaded" (Figure 6.6). Then we must make sure that Fortran types with the `sequence` directive are handled properly by the compiler. Select "Data," and change the "SEQUENCE Types Obey Alignment Rules" setting to "Yes" (Figure 6.7). This setting is equivalent to compiling with the `/align:sequence` option at the command line.

With the Lahey/Fujitsu compiler, these changes are unnecessary; all property settings may be left at their default values.

Now the DLL is ready to be built, using either the "Debug" or "Release" configuration. After building it, you should check the contents of the DLL's export table to make sure the symbolic name is correct. In a command-prompt window, change to the "Debug" or "Release" subdirectory within the `em_exp` project directory, and type this command:

```
dumpbin /exports em_exp.dll
```

The output will look something like this.

```
File Type: DLL

    Section contains the following exports for em_exp.dll

        00000000 characteristics
        40C91357 time date stamp Thu Jun 10 22:05:11 2004
            0.00 version
```

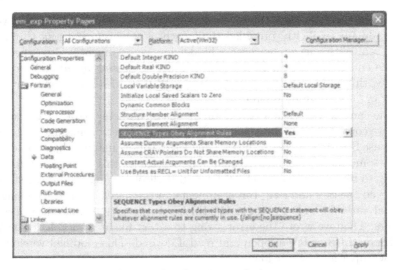

FIGURE 6.7. Setting the alignment rule for sequence types.

```
1 ordinal base
1 number of functions
1 number of names

ordinal hint RVA       name

    1    0 000049D8 em_exponential_
```

Summary

```
1000 .data
4000 .data1
1000 .rdata
1000 .reloc
4000 .text
1000 .trace
```

The most important item here is the name, `em_exponential_`, which is as we intended.

6.5.3 Building with Salford Fortran

When we applied the Salford FTN95 compiler to this example, we encountered a small problem. Salford does an excellent job of creating DLLs that receive and pass numeric arguments to S-PLUS and R, but it cannot pass character strings. The reason is that a Salford-compiled DLL requires an extra argument for each character-string argument defining its length. We have already explicitly provided the length of the argument `msg` through `msg_len`. With Salford, however, a hidden integer argument is also expected at the end of the argument list, and it must be passed *by value*. Unfortunately, this will not be possible because S-PLUS and R pass arguments to Fortran DLLs only by reference.

One can survive without the ability to pass character strings. For example, we could return an integer code rather than the error message text and then use S-PLUS to create the text message. Unfortunately, that would require some changes to our error-handling system. In this example, we found a way to report the error to S-PLUS and R through a back door. First, we removed the `msg` and `msg_len` arguments and replaced them with a single logical variable, `error_occurred`, which returns a `TRUE` value if an error occurs. Then we wrote the error string to a text file at a specific location, which S-PLUS or R can open and read if desired. This modified version of the wrapper is shown below.

```
──────────────────  em_exponential_salford.f90  ──────────────
!#################################################################
subroutine em_exponential(n, y, maxits, eps, pi, lambda_1, lambda_2, &
    iter, converged, loglik, score, hessian, error_occurred )
  use error_handler
  use program_constants
  use em_exponential_engine
  implicit none
  ! declare arguments
  integer, intent(in) :: n
  real(kind=our_dble), intent(in) :: y(n)
  integer, intent(in) :: maxits
  real(kind=our_dble), intent(in) :: eps
  real(kind=our_dble), intent(inout) :: pi, lambda_1, lambda_2
  integer, intent(out) :: iter
  logical, intent(out) :: converged
  real(kind=our_dble), intent(out) :: loglik, score(3), hessian(3,3)
  logical, intent(out) :: error_occurred
  ! declare locals
  integer(kind=our_int) :: ijunk
  type(error_type) :: err
  character(len=512) :: msg
  ! begin
  call err_reset(err)
  ijunk = run_em_exponential( y, pi, lambda_1, lambda_2, &
    iter, converged, loglik, score, hessian, err, maxits, eps )
  if( err_msg_present(err) ) then
    error_occurred = .true.
    call err_get_msgs(err, msg, "UNIX")
    open( 9, file="c:\em_exp_error.txt" )
    write(9,*) msg
    close(9)
  else
    error_occurred = .false.
```

```
      end if
      return
   end subroutine em_exponential
   !###################################################################
```

These code changes for Salford are not necessary if your DLL is to be called by SAS or by another application that can pass extra arguments by value.

The only compilation option required for creating a compatible DLL in Salford is `-export:EM_EXPONENTIAL`, which is given at the command line. This name must be given in all capital letters or S-PLUS and R will not be able to import the symbol. We could not find a way to make Salford apply a trailing underscore to the symbolic name. In the Salford-compiled DLL, the name is simply `EM_EXPONENTIAL`. To call a DLL procedure from S-PLUS or R without the appended underscore in its symbolic name, we were able to use the `.C()` function rather than `.Fortran()`.

6.5.4   Calling the DLL Procedure from S-PLUS and R

Now we demonstrate how to use the DLL from S-PLUS and R. First, we write a function that applies the model-fitting procedure to a sample of data held in a numeric vector.

```
                          em_exponential.ssc
em.exponential <- function( y, start, eps=.00001, maxits=10000 ){
   # client function for the em exponential dll server
   if( missing(start) ){
      # generate starting values by a random split
      w <- sample( 1:2, length(y), replace=T )
      start <- list(
         pi = mean( w==1 ),
         lambda.1 = 1./mean( y[w==1] ),
         lambda.2 = 1./mean( y[w==2] ) ) }
   msg.len <- 255
   msg <- ""
   for(i in 1:msg.len) msg<-paste(msg," ",sep="")
   tmp <- .Fortran("em_exponential",
      n = length(y),
      y = as.double(y),
      maxits = as.integer(maxits),
      eps = as.double(eps),
      pi = as.double(start$pi),
      lambda.1 = as.double(start$lambda.1),
      lambda.2 = as.double(start$lambda.2),
      iter = integer(1),
      converged = logical(1),
      loglik = numeric(1),
      score = numeric(3),
      hessian = matrix(0.,3,3),
      msg.len = as.integer(msg.len),
      msg = msg)
```

```
  est <- list( pi = tmp$pi, lambda.1 = tmp$lambda.1,
     lambda.2 = tmp$lambda.2 )
  list( est = est, converged = tmp$converged )
  msg <- tmp$msg
  if( is.all.white( msg ) )
     msg <- NULL
  else{
     # trim off the white space from msg and print to screen
     i <- msg.len
     while( ( substring( msg, i, i) == " " ) & ( i >= 0 ) ) i <- i-1
     msg <- paste( substring( msg, 1, i ), "\n", sep="" )
     cat(msg)}
  result <- list(
     est = est,
     iter = tmp$iter,
     converged = tmp$converged,
     logliklihood = tmp$loglik,
     score = tmp$score,
     hessian = tmp$hessian,
     msg = msg)
  result}
######################################################################
```

The actual call to `em.exponential` is accomplished with `.Fortran()`.
The first argument to `.Fortran()` is the symbolic name of the exported
procedure *without* the underscore. The remaining arguments are the actual arguments that will be passed to the wrapper. The returned result
`tmp` is an S-PLUS list whose components are the values of the arguments
after the Fortran procedure has run. Naming the arguments in the call to
`.Fortran()` will apply those names to the components of `tmp`. Notice the
use of `as.integer` and `as.double`, which ensure that numeric data are
passed in the form that the DLL is expecting.

The function shown above is also suitable for R, except for one line. The
command `is.all.white` does not exist in R, so the line

```
if( is.all.white( msg ) )
```

may be replaced with

```
if( all(strsplit(msg,"")[[1]]==" ") )
```

Before using this function in the S-PLUS or R session, we need to load
and link the DLL. In S-PLUS, a DLL is loaded using `dyn.open()`; in R,
the corresponding function is `dyn.load()`. This step only needs to be done
once per session. Here is an example of loading and using the DLL on a
small dataset in S-PLUS:

```
dyn.open("em_exp.dll")    # in R, use dyn.load() instead

y <- c( 5.6, 0.7, 2.4, 2.2, 4.5, 0.6, 2.3, 3.1, 1.6, 2.2,
```

```
        0.1, 4.9, 9.0, 7.4, 1.8, 9.7, 0.9, 1.0, 0.7, 3.4,
        1.8, 0.5, 0.1, 0.7, 0.1, 6.6, 1.6, 8.6, 0.3, 0.1,
        4.2, 0.8, 3.1, 0.2, 1.0, 2.0, 2.3, 0.8, 6.6, 1.2,
        0.3, 2.7, 0.5, 0.7, 1.8, 1.5, 2.8, 18.3, 1.2, 0.6)

tmp <- em.exponential(y)
```

For dyn.load() and dyn.open() to succeed, the DLL file must reside
in a directory where S-PLUS or R can find it. In S-PLUS, you can place
it in the current working directory (i.e. the value of the environmental
variable S_PROJ). S-PLUS can also locate the DLL in any directory listed
in the Windows PATH variable. R does not search along the Windows
path, so if you are using R, you need to either place the DLL in the current
working directory or specify a full path for the DLL file in the argument
to dyn.load().

Here we apply the procedure to our data and display the parameter
estimates, the score functions, and the inverse of the information matrix.

```
> y <- c( 5.6, 0.7, 2.4, 2.2, 4.5, 0.6, 2.3, 3.1, 1.6, 2.2,
+          0.1, 4.9, 9.0, 7.4, 1.8, 9.7, 0.9, 1.0, 0.7, 3.4,
+          1.8, 0.5, 0.1, 0.7, 0.1, 6.6, 1.6, 8.6, 0.3, 0.1,
+          4.2, 0.8, 3.1, 0.2, 1.0, 2.0, 2.3, 0.8, 6.6, 1.2,
+          0.3, 2.7, 0.5, 0.7, 1.8, 1.5, 2.8, 18.3, 1.2, 0.6)
>
> tmp <- em.exponential(y)
> print( tmp$est )
$pi
[1] 0.5354197

$lambda.1
[1] 0.6780908

$lambda.2
[1] 0.2379533

> print( tmp$score )
[1]  0.0019629725 -0.0004596604 -0.0008722596
> print( solve( -tmp$hess ) )
             [,1]        [,2]        [,3]
[1,]  0.19482037 -0.14959040 -0.04188542
[2,] -0.14959040  0.16424997  0.02885927
[3,] -0.04188542  0.02885927  0.01226698
```

6.5.5 Calling the Function from SAS/IML

In SAS, it is easy to call an external DLL from PROC IML. IML invokes
DLL procedures through the routine **modulei**, as shown in this example:

```
―――――――――――――――――――――― em_exponential.sas ――――――――――――
proc iml;
  /*  IML module for exponential mixture model */
  start em_exp(y, eps, maxits, pi, lambda_1, lambda_2,
    score, hessian);

    filename sascbtbl 'em_exp.cbt';

    /* initializations */
    n = nrow(y);
    converged = 0;
    iter = 0;
    loglik = 0.0;

    /* Make enough room for 255 characters in msg */
    msg = ' ';
    do i=1 to 51;
      msg = msg + '     ';
    end;
    blank_msg = msg;

    if type(pi) = 'U' then do; /* pi not defined */
      /* generate starting values by a random split */
      w = round(uniform(shape( 0, 1, n ))) + 1;
      pi = sum( w=1 )/n;
      lambda_1 = sum(w=1)/( (w=1) * y );
      lambda_2 = sum(w=2)/( (w=2) * y );
    end;

    /* Shape the score and Hessian arrays */
    score = shape( 0.0, 3, 1);
    hessian = shape( 0.0, 3, 3);

     /* Run estimation */
    call modulei('em_exponential_', n, y,
      maxits, eps, pi, lambda_1, lambda_2,
      iter, converged, loglik, score, hessian,
      255, msg);

    if msg ^= blank_msg then do;
      print 'Error: ', msg;
    end;
  finish; /* em_exp */

  /*  good sample */
  y = { 5.6, 0.7, 2.4, 2.2, 4.5, 0.6, 2.3, 3.1, 1.6, 2.2,
        0.1, 4.9, 9.0, 7.4, 1.8, 9.7, 0.9, 1.0, 0.7, 3.4,
        1.8, 0.5, 0.1, 0.7, 0.1, 6.6, 1.6, 8.6, 0.3, 0.1,
        4.2, 0.8, 3.1, 0.2, 1.0, 2.0, 2.3, 0.8, 6.6, 1.2,
```

```
           0.3, 2.7, 0.5, 0.7, 1.8, 1.5, 2.8, 18.3, 1.2, 0.6 };

   maxits = 10000;
   eps = 1.0E-5;

    run em_exp(y, eps, maxits, pi2, lambda_1, lambda_2,
      score, hessian);

   /* Display results */

   print 'Final Estimates:';
   print (pi2), , (lambda_1), , (lambda_2);
   print (score);
   cov = inv(-hessian);
   print (cov);

quit;
```

This code requires an accompanying text file, `em_exp.cbt`, which defines
the DLL calling conventions. The format of this file is described in the
SAS Companion for the Microsoft Windows Environment, Version 8 (SAS
Institute, 2000). Our version of the file looks like this:

```
—————————————————————— em_exp.cbt ——————

routine em_exponential_
  module=em_exp
  transpose=yes;
arg 1 num input byaddr format=ib4.;
arg 2 num input byaddr format=rb8.;
arg 3 num input byaddr format=ib4.;
arg 4 num input byaddr format=rb8.;
arg 5 num update byaddr format=rb8.;
arg 6 num update byaddr format=rb8.;
arg 7 num update byaddr format=rb8.;
arg 8 num output byaddr format=ib4.;
arg 9 num output byaddr format=ib2.;
arg 10 num output byaddr format=rb8.;
arg 11 num update byaddr format=rb8.;
arg 12 num update byaddr format=rb8.;
arg 13 num input byaddr format=ib4.;
arg 14 char update byaddr format=$CHAR255.;
```

The first line of this file contains the export's symbolic name *including*
the underscore, if present. The second line, `module=em_exp`, provides the
name of the DLL file without the `.dll` extension. The line `transpose=yes`
is necessary because normally SAS/IML would pass the arrays in row-
major order, whereas Fortran expects array arguments to be passed in
column-major order. The remaining lines give details for each of the procedure's arguments—the type of data, the argument's intent, the passing
method (`byaddr` means by reference), and the type of data contained in

the argument. The `.cbt` file can be located anywhere, provided that the SAS program gives the full file path in the `filename sascbtbl` statement. Alternatively, you may place the file in the same folder as the `*.sas` file, as we have done in this example.

Once again, it is important to place the DLL file in the proper location. For SAS, you may place it in the `sasexe` subdirectory of the SAS installation directory. Alternatively, you may place it in the same directory as the `*.sas` file, above, or in any location specified by the system's `PATH` environment variable.

The output from this SAS program will be similar to the following:

```
─────────────────── em_exponential.lst ───────────────────

                       Final Estimates:


                            PI2

                         0.5363882

                         LAMBDA_1

                         0.6773083

                         LAMBDA_2

                         0.2377425


                           SCORE

                        -0.001932
                        0.0004536
                        0.000858


                            COV

           0.195054 -0.149587 -0.041936
          -0.149587 0.1638826 0.0288602
          -0.041936 0.0288602 0.0122806
```

6.6 Shared Objects in Unix and Linux

Shared objects in Unix and Linux systems are analogous to DLLs in Windows. Fortran compilers on Unix and Linux systems can produce shared object files containing procedures that may be called from programs written in other languages. As with DLLs, shared objects are compiled and linked from source code and are linked and loaded dynamically with another program at run time. All of the technical issues regarding calling conventions and argument passing that we have presented regarding Windows DLLs also apply to shared objects.

Shared objects by convention have the file extension .so, and often it is helpful to give the file a name beginning with "lib". For example, you may create a shared object called libMyStats.so. To create a shared object, one generally must compile source-code files into object code (*.o) files and then link them, specifying an option indicating that the product is a shared object. For specific details about creating shared objects, we recommend that the reader refer to the compiler's documentation.

Distributing shared objects to other users may require special treatment during compilation. Additional run-time libraries, which accompany the particular Fortran compiler, may be required in order to run the code on another user's machine. Distribution of run-time libraries must be done in accordance with the license agreement that accompanies the compiler. For details about distributing compiled shared objects in Linux, we recommend that you consult the compiler's documentation.

6.6.1 An Example: Extending S-Plus and R via a Fortran Shared Object in Linux

This example will use the Intel Fortran compiler for the Linux operating system. First, we place the following files in a directory on a Linux system with this compiler:

```
constants.f90
error_handler.f90
em_exponential_engine.f90
em_exponential.f90
```

These source files are unchanged from previous examples, with one exception. In the file em_exponential.f90, we must remove the line containing the !DEC$ ATTRIBUTES directive. With shared objects, all symbols are exported by default.

Creating a shared-object library requires two steps. First, we compile the source files into static-object files. The alignment of sequence types is handled appropriately using the -align sequence option.

```
ifort -c -align sequence constants.f90
ifort -c -align sequence error_handler.f90
ifort -c -align sequence em_exponential_engine.f90
ifort -c -align sequence em_exponential.f90
```

The next step is to link the static objects into a shared-object file. This is accomplished with the -shared option. We specify that the output will be named em_exponential.so.

```
ifort -shared *.o -oem_exponential.so
```

To test the shared object, we run the same R code as before except for one minor change:

```
dyn.load("em_exponential.dll")
```

must be changed to

```
dyn.load("em_exponential.so").
```

When we execute this code using R in the Linux environment, it produces the expected output.

7
Creating COM Servers

Thus far, we have discussed how to create Fortran modules that perform computational tasks and shown how to use those modules in console applications and DLLs. Many of today's computer users, however, are unaccustomed to obtaining and using software in those forms. They have come to expect the convenience of user-friendly graphical interfaces (GUIs) or easily installable plug-ins for their favorite applications. Modern programs with GUIs and plug-ins are built from software components. These components are object-oriented and contain object classes, properties and methods. Like the conventional DLLs discussed in Chapter 6, software components are precompiled into binary code and can be shared by multiple applications. But they are more intelligent and versatile than conventional DLLs and can be used more easily because their calling conventions have been standardized.

This chapter focuses on the Component Object Model (COM), the predominant standard for Windows operating systems. We explain the essential features of COM and demonstrate how to create components that adhere to the COM standard. These components are called COM servers. Along the way, we describe some new tools we have created for the Intel Visual Fortran compiler to automatically wrap a Fortran module with the code necessary to convert it to a COM server. If you have written a module using the pseudo object-oriented style described in Chapters 3 and 5, this chapter will show you how to turn it into a COM server.

7.1   A Simple Example

7.1.1   The magic8 Fortran Module

As a first demonstration of a COM server, we have chosen a simple application of a random number generator: the Magic 8-ball®. This classic American toy has been manufactured and sold for more than fifty years. It is a hollow replica of a black 8-ball used in pool or billiards with a small window. Inside the ball, floating in a mysterious cloudy liquid, is an icosahedron (a 20-sided die with triangular faces) inscribed with various messages. The user asks the ball a question, shakes it, and waits for a message to appear in the window. Ten of the possible responses are positive ("Yes"; "Signs point to yes"; "Without a doubt"; "As I see it, yes"; "You may rely on it"; "It is decidedly so"; "Most likely"; "Outlook good"; "Yes—definitely"; "It is certain"); five are negative ("My sources say no"; "Outlook not so good"; "My reply is no"; "Don't count on it"; "Very doubtful"); and five are vague ("Concentrate and ask again"; "Reply hazy, try again"; "Better not tell you now"; "Cannot predict now"; "Ask again later"). The behavior of the Magic 8-Ball can be easily simulated by generating a uniform random number, mapping it to the integers from 1 through 20, and returning a character string containing the corresponding message. Here is a Fortran module that mimics the Magic 8-Ball.

```
———————————— magic8.f90 ————————————
!######################################################################
module magic8
   implicit none
   private ! by default
   public :: get_magic8_msg
   character(len=*), parameter :: modname = "magic8"
contains
   !###################################################################
   integer function get_magic8_msg( msg ) result(answer)
      implicit none
      ! Generates a random message from the Magic 8-ball. The argument
      ! msg should be at least 26 characters long, otherwise the
      ! message may be truncated.
      !
      ! Because this uses the Fortran intrinsic generator, you should
      ! set the seed (e.g. by calling random_seed) before using this
      ! function.
      character(len=*), intent(out) :: msg
      ! locals
      real :: u
      integer :: i
      ! generate a random integer between 1 and 20
      call random_number(u)
      i = 20*u + 1
      select case( i )
         ! positive responses
```

```
          case( 1)
             msg = "Yes."
          case( 2)
             msg = "Signs point to yes."
          case( 3)
             msg = "Without a doubt."
          case( 4)
             msg = "As I see it, yes."
          case( 5)
             msg = "You may rely on it."
          case( 6)
             msg = "It is decidedly so."
          case( 7)
             msg = "Most likely."
          case( 8)
             msg = "Outlook good."
          case( 9)
             msg = "Yes - definitely."
          case(10)
             msg = "It is certain."
          ! negative responses
          case(11)
             msg = "My sources say no."
          case(12)
             msg = "Outlook not so good."
          case(13)
             msg = "My reply is no."
          case(14)
             msg = "Don't count on it."
          case(15)
             msg = "Very doubtful."
          ! vague responses
          case(16)
             msg = "Concentrate and ask again."
          case(17)
             msg = "Reply hazy, try again."
          case(18)
             msg = "Better not tell you now."
          case(19)
             msg = "Cannot predict now."
          case default
             msg = "Ask again later."
          end select
          ! normal exit
          answer = 0
          return
       end function get_magic8_msg
    !###################################################################
end module magic8
!###################################################################
```

This simple module follows the principles set forth in previous chapters for pseudo object-oriented programming in Fortran. The public interface of this module consists of the single function `get_magic8_msg`. The name of this function indicates that it is a get method for the property `msg`. There is no corresponding put method, so `msg` is a read-only property. A Fortran program or routine that uses this module can invoke the public function `get_magic8_msg` and retrieve a random message via its `msg` argument, which has `intent(out)`.

7.1.2   The Magic8 COM Server

Invoking the get method for `msg` from a Fortran program is a simple matter. It becomes equally straightforward to invoke this method from other Windows applications when we package the module as a COM server. Skipping over the details for now, we have already created a COM server called Magic8 and placed it on our Web site.

The Magic8 COM server consists of a single file called `magic8_com.exe`. The `.exe` suffix suggests that this is a program that can be run by typing the filename at a command prompt or by double-clicking on the file's icon in a file-management program such as Windows Explorer. This is not an ordinary program, however; it's a precompiled software module that defines an object class with a single property and method. You can download the Magic8 COM server, install it on your computer, and use it in a variety of ways, as we now describe.

7.1.3   Installing the Magic8 COM Server

Where should you put the file `magic8_com.exe`? If this were an ordinary executable program invoked from a command prompt, it would need to reside in the current working directory or somewhere along the Windows search path (e.g., as specified by the `PATH` environment variable). Because this is a COM server, however, you may place it almost anywhere on your Windows file system. Other programs will be able to find it once it has been registered with Windows.

To install the Magic8 COM server on your computer, download the file `magic8_com.exe` to any convenient location on your computer's hard drive. (You may wish to create a directory called `C:\servers` and place the file there.) Then register the COM server in the following way. First, open a command-prompt window. Next, change the working directory to the directory where `magic8_com.exe` is located (for example, by typing `cd C:\servers`). Finally, type

```
magic8_com /regserver
```

at the command prompt to register the server. The file `magic8_com.exe` is not an installer program; it is the COM server itself. When using a COM server, one would not normally run it from a command prompt. We are doing so at this time merely to register it with Windows.

The Magic8 COM server is now installed, registered, and ready to speak. Any application that uses a COM server is called a COM client. In the next few subsections, we show how to create simple clients to make the 8-Ball speak in Excel, S-PLUS, R, MATLAB, and SAS. Whether or not you use all of these software packages, it will be instructive to look over these code samples in order to understand how COM servers are used in different client environments.

7.1.4   The 8-Ball Speaks in Excel

Microsoft Excel is widely used for organizing data and performing basic calculations. Advanced users of Excel know how to automate tasks by writing macros. An Excel macro is a script written in a language called Visual Basic for Applications (VBA). If you have Excel installed on your computer, then you also have VBA. Only a few lines of VBA code are needed to invoke the Magic 8-Ball and make it speak.

To enter the necessary VBA code, first launch the Excel program. Then choose from the menu Tools → Macro → Visual Basic Editor, and the Microsoft Visual Basic editor window will then appear. In the editor, under "Microsoft Excel Objects," double-click on Sheet1. This will open a window in which VBA code may be added behind the current spreadsheet.

If the Magic8 COM server has been registered with Windows, then the VBA editor already knows about it, and you can add a reference to the Magic8 COM server in this VBA application. To do this, go to the Microsoft Visual Basic menu and choose Tools → References . . ., and a dialog box will appear. In this dialog box, you should be able to scroll down and find "Magic8Obj 1.0 Type Library." Click the checkbox next to it, and then click OK. Then add the following code in the editor window.

```
Sub speak()
    ' VBA script that uses the Magic8Obj COM Server

    ' declare object
    dim objMagic8 as Magic8Obj

    ' create object instance
    objMagic8 = createObject("magic8_com.Magic8Obj")

    ' invoke 'msg' property and display in message box
    call MsgBox(objMagic8.msg)

    ' release object from memory
    Set objMagic8 = Nothing
End Sub
```

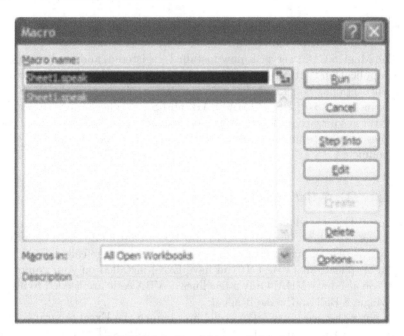

FIGURE 7.1. Excel Macro dialog box.

Let's examine this VBA code in detail. It defines a single subroutine called `speak()`. The lines that begin with a single quote (') are comment lines. The first executable line, which begins with `dim`, is a dimensioning statement; it declares the variable `objMagic8` to be a pointer to a `Magic8Obj` object. The next executable line creates an instance of the object. (Later, we will discuss in detail what the object is. For now, just consider `createObject` to be the command that gets the Magic8 COM server up and running.) The string argument `"magic8_com.Magic8Obj"` identifies the Magic8 COM server's object. The portion of this string before the period (`"magic8_com"`) is the name of the COM server, and the portion after the period (`"Magic8Obj"`) is the name of the object class. The instance of the object is called `objMagic8`, named after the pointer that has been assigned to it.

The next executable statement does two things. The expression within parentheses (`objMagic8.msg`) invokes the get method for the `msg` property and evaluates to a character string. The `call MsgBox` command then takes that character-string message and displays it in a Windows message box (i.e., a dialog box window).

Now that we understand what the VBA code does, let us return to the main spreadsheet window of Excel to run the code. From the Excel menu, choose Tools → Macro → Macros..., and a dialog box will appear, as shown in Figure 7.1. Move your mouse cursor to this box and press Run. Another

FIGURE 7.2. Dialog box generated by the Excel macro showing the Magic 8-Ball's response.

dialog box will appear as shown in Figure 7.2, displaying a message from the Magic 8-Ball.

7.1.5   The 8-Ball Speaks in S-PLUS and R

S-PLUS for Windows is also designed to interact with COM servers. In S-PLUS, one can write either a COM client or a COM server. Here we show how to write a client for the Magic8 server.

Open S-PLUS for Windows and choose from the menu File → New...; in the resulting dialog box, choose Script File and press OK. This creates a new script file. In the top portion of the script window, enter the following S language commands.

```
# S-Plus script that uses the Magic8Obj COM Server

# create object instance
objMagic8 <- create.ole.object("magic8_com.Magic8Obj");

# invoke 'msg' property
msg <- get.ole.property(objMagic8,"msg" );

# release object from memory
release.ole.object(objMagic8);

# print message
cat(paste("\n\n", msg, "\n"));
```

Before executing the script file, save it under the name magic8. Then execute the script by choosing from the menu Script → Run or by simply pressing the "run" button ▶ or the F10 keyboard key. After echoing the script commands, the output window will show the 8-ball's reply:

    "Yes - definitely."

Interoperability with COM is not a standard part of the R programming environment. However, a COM interface for R on the Windows platform (rcom by Thomas Baier) is available for download. A link to rcom can be found on the Web site for this book. A Magic8 client for R is shown below.

```
# R script that uses the Magic8Obj COM Server

# Load rcom library
library(rcom);

# create object instance
objMagic8 <- comCreateObject("magic8_com.Magic8Obj");

# invoke 'msg' property
msg <- comGetProperty(objMagic8,"msg" );

# release object from memory
rm(objMagic8);

# print message
cat(paste("\n\n", msg, "\n"));
```

Notice that the code for R is almost identical to that from S-PLUS except for a few differences in the command names. Another difference is that, in order to release the object, we simply remove the pointer with the rm command. (This is a feature of the "garbage collection" system that R uses to manage its memory. When we remove the pointer, the instance of the object is left in memory with no way of accessing it. Fortunately, R will collect the garbage—that is, R automatically deallocates sections of memory that have become inaccessible or unusable.)

7.1.6   The 8-Ball Speaks in MATLAB

MATLAB is popular among scientists and engineers for its ability to manipulate matrices. The Windows version of MATLAB supports COM and can be used either as a server or a client. Our MATLAB Magic8 client is quite similar to the clients for S-PLUS and R. To create this client, start MATLAB and select from the menu File → New M − file. Then enter the following code in the code editor.

```
% Matlab M file that uses the Magic8Obj COM Server

% create object instance
objMagic8 = actxserver( 'magic8_com.Magic8Obj');

% invoke 'msg' property
msg = get(objMagic8,'msg');

% release object from memory
release(objMagic8);
```

```
% print message
fprintf('\n\n %s\n',msg);
```

Save this file under the name `magic8.m`. To run the client, go to the code editor menu and select Debug → Run, or simply press the "run" button ⏵ or the F10 keyboard key. The 8-ball's reply will appear in the MATLAB command window:

```
Better not tell you now.
```

7.1.7 The 8-Ball Speaks in SAS

SAS was not designed to support COM. However, we have created a tool called SASCOMIO (for "SAS/COM interoperability") that allows COM servers to be accessed from regular SAS programs. SASCOMIO, along with its documentation and installation instructions, can be downloaded from the Web site that accompanies this book. (The SASCOMIO tool is actually a Fortran DLL. For those who are interested, the source code for this DLL is also provided on our Web site.)

Below is a COM client for the Magic8 COM server that uses SASCOMIO. Notice the extra step required to load our SAS/COM interoperability function definitions. This and other issues of using COM with SAS will be discussed in Chapter 8.

```
proc iml;
    /* SAS/IML program that uses the Magic8Obj COM Server */

    /* initializations */
    result = 0;
    msg = "12345678901234567890123456"; /* 26 characters */

    /* load SAS/COM interop function definitions */
    filename sascbtbl 'e:\project\magic8_com\clients\sascomio.cbt';

    /* create object instance */
    pMagic8 = modulein('createobj', 'magic8_com.Magic8Obj' );

    /* invoke 'msg' property */
    result = modulein('getpropchar', pMagic8, 'msg', 1, 1, msg, 0);

    /* release object from memory */
    call modulei('releaseobj', pMagic8);

    /* print message */
    print msg;
quit;
```

The output from SAS after running this code looks like this:

```
MSG
My sources say no.
```

7.1.8   Exercises

1. Download the Magic8 COM server. Place the file in an appropriate location and register it with the Windows operating system.

2. Create a PowerPoint presentation that uses the Magic8 COM server. Create a slide with an action button that, when pressed, runs a VBA macro and displays a dialog box that shows the Magic 8-Ball's answer.

7.2   COM Server Basics

7.2.1   References on COM

The ability to call the same software component from such disparate environments as Excel, S-PLUS, R, MATLAB, and SAS makes COM seem almost magical. The technology behind this magic is actually quite sophisticated. A detailed exposition of COM lies well beyond the scope of this book. For advanced programmers who wish to learn about it, many resources are available. The official COM specification is available from Microsoft at this Web address:

```
http://www.microsoft.com/COM
```

We also recommend *The Essence of COM* by Platt (2000). Although that book focuses on programming in C and C++, the concepts of COM are the same regardless of what language is used. Other useful information pertaining to COM and Fortran can be found in the documentation accompanying the Intel Visual Fortran compiler. Look in the manual *Intel(R) Fortran Libraries Reference*, in the "COM and AUTO Routines" section, to learn about the Intel Visual Fortran COM libraries IFCOM and IFAUTO.

Technicalities aside, COM can be viewed rather simply as an extension of the object-oriented concepts introduced in Chapter 3. In particular, before delving into the discussion below, make sure that you understand the basic ideas that were covered in Section 3.1.

7.2.2   COM, Windows, and .NET

Just as integrated circuit chips are the building blocks of modern computer hardware, software components are the building blocks of modern computer software. Early programmers for Windows envisioned that applications would eventually be assembled from specialized components. For example, a text-editing component, a typesetting component, and graphical display component, along with a spell-checker, dictionary, and thesaurus, could be assembled to create a sophisticated word processor. Through object-oriented component technologies such as COM, much of this vision has become reality.

COM is now an integral part of 32-bit Microsoft Windows operating systems. Much of Windows is built on COM. Although the Windows platform is continually evolving, component technologies are remaining as the dominant feature of Windows architecture. Software components are finding their way into other operating systems as well.

If you have taken a look at recent software development literature, you may have seen little about COM but a lot about a newer technology called .NET (pronounced "dot-net"). The .NET framework, like COM, promotes the use of object-oriented software components. In fact, .NET was designed to interact with COM, as we shall demonstrate in the next chapter. The basic principles established by COM—including the interface, methods, properties, servers, and clients—have been implemented and extended by .NET. Countless books and articles are now available about .NET, and you can learn much about the latest concepts of software component development from those resources.

Two Fortran compilers, Salford and Lahey/Fujitsu, now allow programmers to develop components and applications for .NET. These tools show great promise for developing components as they evolve along with the .NET framework. This book, however, will focus specifically on COM servers.

7.2.3   COM Servers versus Conventional DLLs

After developing a COM server in a given programming language, the programmer must compile and link it. Afterward, the COM server may be used in other projects written in any of a number of languages. In other words, the language used to develop COM servers and the applications that use them is unimportant.

Multilanguage programming is possible with conventional DLLs as described in Chapter 6. However, the use of conventional DLLs is complicated by the lack of standard calling conventions. As a result, DLLs must often be customized for each application, and the types of data that may be passed as arguments across languages can be quite limited. But COM servers (and clients) have a standard mechanism for invoking routines and passing arguments. This is because COM is a standard that defines the interface between the server and the application that uses it (the client).

Another way in which COM servers differ from conventional DLLs is that they are object-oriented. After all, "object" is COM's middle name. DLLs are simply a way of packaging computational procedures. But COM is a more powerful paradigm that allows you to create instances of objects and put and get properties and to invoke methods. COM allows you to initialize the data when an instance is created and to clean up afterward when the instance is destroyed.

The COM standard also eliminates "DLL hell." COM servers are independently installed and registered in the operating system, and each provides its own internal documentation. By nature they are language-independent. And they have unique identifications beyond their filenames. That is, two COM servers may have identical filenames but still be uniquely distinguished. COM even provides unique identifications for different versions of the same server.

Some of these advantages were already evident in our Magic 8-Ball example. Our Magic8 COM server's `msg` function was invoked from five client environments without alteration, using only a few lines of code, and the form of the client code was similar among the various applications. No concern was paid to calling conventions; the inner workings of COM made sure that data were passed between the client and server correctly.

On the other hand, COM servers do have a couple of disadvantages relative to DLLs. First, it takes a fair amount of extra code to produce a COM server from a Fortran module. The extra code is not standard Fortran 95 but relies on vendor-specific libraries and the Microsoft Interface Definition Language. But this code follows a fixed pattern and can be generated automatically. If you are using Intel Visual Fortran and have coded your Fortran module in the pseudo object-oriented style described in Chapters 3 and 5, we provide a script that will automatically wrap your module with all the extra code you need to turn it into a COM server.

A second disadvantage of COM servers is the brief delay that occurs at the beginning of procedure calls. A conventional DLL is like any other part of the running program: routines within the DLL communicate with other parts of the program without delay. Communication between a client and a COM server, however, must be mediated by the COM infrastructure in the Windows operating system, which takes some time. Under most circumstances, the time delay is minuscule. But if a COM server's methods or properties are invoked numerous times in rapid succession—if the client uses a loop, for example—the cumulative effect can be noticeable. If speed is critical and you expect your Fortran procedures to be called iteratively, then you may want to package them in a conventional DLL. On the other hand, you might also consider changing the project's overall design so that the loop is moved out of the client and placed inside the Fortran code.

7.2.4   The Object-Oriented Contract

The COM standard defines the rules of communication between a COM server and its client. These rules are often called a contract. The server and client are both expected to carefully adhere to these rules. Details of the contract define protocols for invoking properties and methods and formats for data transfer.

Upon request from a COM client, a COM server may create an instance of an object. When a new object is created, the server gives the client a pointer to the object. The pointer is simply a name or a handle by which the client can subsequently refer to the object. Using this pointer, the client can directly invoke any of the object's properties and methods. The client can also release the object from memory when its life cycle is over. This standard object-oriented approach allows clients to use the server without detailed knowledge of its internal workings.

To make these ideas more concrete, suppose that we have a COM server for performing linear regression analysis. This server defines an object class with properties and methods for putting the data (i.e., the response variable and predictors), computing the ordinary least-squares solution, and retrieving the estimated coefficients, residuals, and so on. A COM client may create one instance of this object (instance A) into which it loads one set of data. It may also create a second instance (instance B) into which it loads an entirely different set of data. Suppose that the method for computing the least-squares solution is called `fit`. When the client invokes `fit`, how does the server know which dataset to use? The client applies the method to a particular instance by invoking `A.fit` or `B.fit`.

An active COM server resides in sections of memory. One section holds a single copy of the binary code that implements the object class definitions along with their properties and methods and any common data. The class definitions, properties, and methods are called the *interface*. Whenever a new object is instantiated, another section of memory is added to hold the data unique to that object. The latter is called *per-instance data*. The per-instance data for an object may grow or shrink over time as the object's properties change. When a client is finished with the object and disposes of it, its per-instance data are released and that portion of memory returns to the heap. Finally, when all objects are destroyed, the binary code for the COM server and any common data are removed.

A conceptual diagram that depicts an object, its interface, and per-instance data is shown in Figure 7.3. The interface is the sole path through which a client may interact with the object. Two-way arrows represent read-write properties, and the one-way arrows represent read-only properties. The per-instance data are encapsulated within the object. The interface is distinct from the object, representing that contract of communication to which both the client and the server are bound.

7.2.5 In-Process versus Out-of-Process Servers

COM servers come in two different flavors: in-process and out-of-process. An in-process server is packaged as a DLL, whereas an out-of-process server is packaged as an executable (`.exe`) file. When creating a COM server, one

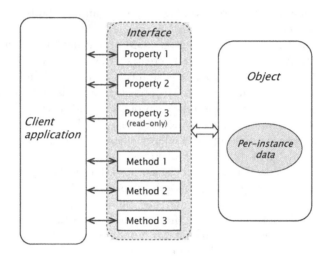

FIGURE 7.3. Conceptual diagram of an object and its interface.

must choose one kind or the other, so let us briefly consider the differences between them.

When a client application (e.g., Microsoft Excel) is running, the client program is loaded into a section of computer memory along with all of its associated DLLs. These DLLs are dynamically loaded into the same virtual memory space as the program itself (Section 6.3). Now suppose that we want to use a COM server (e.g., Magic8) from the client. If it were packaged as an in-process server, it would be treated like any other DLL being used by the client. Its code would be dynamically linked and loaded into the client's virtual memory space. The COM server's instructions and data would act as if they were part of the client program.

If, on the other hand, we were using an out-of-process COM server, the Windows operating system would launch the server as a stand-alone program with its own memory space. The COM server would assume responsibility for creating and managing the instances of objects it spawns. Because the client and server are running as separate processes, the Windows operating system must negotiate communication between them.

Whether the COM server is running in the client's process or in its own, it will produce exactly the same results. Whether one decides to make an in-process or an out-of-process server, the Fortran code inside is identical; only the build procedure is different. The choice between an in-process or out-of-process server is primarily a matter of personal taste. However, there are two minor issues worth considering. The first issue pertains to what happens in the event of a crash. If a COM server unexpectedly terminates due to an arithmetic exception or some other error, and if the server is running in the same process as the client, then the client application will probably terminate as well. If the server is running in its own process apart from the client, however, unexpected termination of the server may cause the client to issue an error message, but the client will keep running.

The second issue concerns the brief time delay that arises whenever a COM server's properties or methods are invoked. This delay reflects the overhead cost of transferring instructions and data between client and server through the COM interface. If the server lives in the same virtual memory space as the client, the delay may be slightly shorter than if it is running in a separate process.

The name of our Magic8 COM server file (`magic8_com.exe`) shows that it is an out-of-process server. The remainder of this chapter will deal primarily with out-of-process COM servers.

7.3   Example: A COM Server for Logistic Regression

7.3.1   Producing COM Servers with Intel Visual Fortran

At the time this book is being published, Intel does not officially advertise the building of COM servers as one of the features of its Fortran compiler. Nevertheless, Intel Visual Fortran does provide libraries capable of doing this. These libraries have been held over from the time when the product was owned by Compaq. The Professional Edition of Compaq Fortran had a tool called *COM Server Wizard* that guided users through the process of building a COM server. Unfortunately, that tool did not survive the transition to Intel Visual Fortran 8.0, and the old Compaq compiler is no longer being sold. With the permission of Intel, however, we are providing our readers with our own tool based on the COM Server Wizard that can create a server from an existing Fortran module, provided that the module is written in the pseudo object-oriented style illustrated in Chapters 3–5.

Our tool for creating COM servers works as follows. First, it scans your Fortran module for properties, methods, and per-instance data types. Next, it automatically generates code files that wrap around your Fortran module and, together with them, can be compiled and linked by Intel Visual Fortran and Microsoft's Visual Studio into a COM server. We now lead you through this process using the ELOGIT example program developed in Chapter 5.

7.3.2   Getting Ready

The ELOGIT program from Chapter 5 was written entirely in standard Fortran 95. Although the program was invoked from a command line and accepted input from a control file, only 10% of its source-code lines had anything to do with reading information from the control file or writing to the screen. Stripping away those lines, the remaining 90%—in particular, the main module elogit_types—stands as an object-oriented component that we can readily turn into a COM server.

Before you can build the ELOGIT COM server, you must perform a few preliminary tasks. First of all, you must download and install ActivePerl™ (version 5.6.1 or later), which is available from the ActiveState Web site:

    http://www.activestate.com

Our code-generating tools are written in the Perl language, so to run them you will need a Perl interpreter. We recommend ActiveState's free implementation called ActivePerl, but other versions of Perl may also work.

Next, you will need our COM Server Generator package, which is available for download from the Web site that accompanies this book. On our home page, look for the "COM Server Generator" link. It will guide you to installation instructions, updated documentation, and a link for downloading the installer. It will also provide links to any additional components that may be required.

7.3.3 Naming the Server and the Class

Every COM server must have a name, and any object class defined by a COM server must also have a name. For example, the Magic8 COM server was named magic8_com, and the class was named magic8obj. For want of better terms, we will call these the servername and classname, respectively.

The servername we have chosen for the ELOGIT example is ELogitSrvr. This will be the project's name in Visual Studio, as well as the name of the server's executable file.

The classname is equally important because COM uses both the servername and the classname when creating an object. In each of the Magic8 clients shown in Section 7.1, the object created by the COM server was identified as *servername.classname*. One important caveat is that you should not choose the types module prefix (for example, elogit) as the classname. The problem is that one of the modules that is automatically generated by our Perl script is named *classname*_types. In this case, the resulting classname would be elogit_types, which conflicts with the name of our primary Fortran module. To avoid this potential conflict, we will apply the classname ELogitObj.

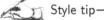 Style tip───

Use Srvr in a servername to distinguish the COM server from a console application. Use Obj in a classname to avoid a naming conflict with the Fortran module itself.

───

7.3.4 Fortran Style Conventions

Our Perl script for generating COM server code assumes that your main Fortran module has been written in the same style as the `elogit_types` module in Chapter 5. Specifically, we require the following.

- Your module must use a `program_constants` module similar to ours.

- The module must use our `error_handler` and `dynamic_allocation` modules.

- All data must be held within instances of a public derived type or types defined in the module. These types may be nested (for example, as in the `elogit_session` type).

- All properties and methods must be defined as public functions of type `integer(kind=our_int)`. Each of these functions must return a value of `RETURN_SUCCESS` or `RETURN_FAIL`. These are Fortran parameters whose values are set in the module `program_constants`.

- A put property function must have a name that follows this form:

  put_*modulename_propertyname*

  The *propertyname* portion may contain embedded underscore (_) characters. Similarly, a get property function must be named:

  get_*modulename_propertyname*

  Any other methods (e.g., computational procedures) must be functions whose names are:

  run_*modulename_methodname*

- The first dummy argument of any put or get function must be the quantity being put or gotten. The next argument(s) must be the public type(s) being passed, followed by the error type (which must be named `err`), followed by any other arguments.

- The argument list for any other method (i.e., a function whose name begins with `run_`) must begin with the public type or types being passed, followed by the `error_type err`, followed by any other arguments being passed to the method. You may also have an additional, optional error type called `warn` for reporting nonfatal problems or other text messages.

The best way to ensure your project conforms to these conventions is to closely follow the programming style of the `elogit_types` module from Chapter 5.

7.3.5 Automatically Generating the COM Server Code

If your Fortran project conforms to these conventions, your source code can be processed by our COM Server Generator to automatically generate additional code files required by Intel Visual Fortran to produce a COM server. To see how this is done, follow these steps with our ELOGIT code.

First, in a convenient location on your computer's hard drive, create a new directory called **ELogitSrvr**. Copy the following ELOGIT source-code files into that directory:

```
constants.f90
elogit_types.f90
elogit_puts.f90
elogit_gets.f90
elogit_nullify.f90
elogit_modelfit.f90
elogit_io.f90
wls.f90
matrix.f90
error_handler.f90
dynalloc.f90
```

Next, open a command prompt window and change the current directory to **ELogitSrvr**. For example:

```
cd c:\project\ELogitSrvr
```

Now run the Perl script to generate the COM server code. The script can be executed from a command prompt window by issuing a command with the following form:

```
perl comserver.pl <servername> <Fortran file>
```

In this command, *servername* is the name of the COM server you want to create, and *Fortran file* is the name of the main pseudo object-oriented types module in your project, the module that defines the public derived types for holding data and contains all the public methods that operate upon these types. In the case of ELOGIT, the command would be:

```
perl comserver.pl ELogitSrvr elogit_types.f90
```

Because this is a new project, the Perl script will prompt you for further information. These prompts and the proper responses to them are shown below.

```
Enter the server's class name: ELogitObj

   1.  In-process (DLL)
   2.  Out-of-process (EXE)
Choose the type of server to create: 2
```

```
Enter the server's version number: 1.0

Enter a help string for the server's object:
Performs logistic regression by weighted least squares

Creating "ELogitSrvrTemplates" directory...
Generating initial hierarchy file...
Scanning "elogit_types.f90" ...
Writing object hierarchy ...
Creating user instance and interface files:
   UIElogitObj.f90
   UElogitObjTY.f90
Generating files from templates:
   server.idl
   ELogitSrvrGlobal.f90
   serverhelper.f90
   exemain.f90
   clsfactty.f90
   clsfact.f90
   ELogitObjTY.f90
   IElogitObj.f90
Copying files:
   variant_conv.f90
   server.rc
Done.
```

(Note: If the in-process server (DLL) option is chosen, the file `exemain.f90` will not be generated, but two other files, `server.def` and `dllmain.f90`, will appear.)

The first time the Perl script is run on a new project, it creates a subdirectory called *servername* Templates. This directory will be used for any subsequent runs of `comserver.pl`. It contains the templates used to automatically generate files and a file called *servername*.hie (the hierarchy file) that contains information about the COM server. From this moment onward, whenever changes are made to public properties and methods in the Fortran main types module, you must rerun the Perl script. For example, if you add a new public property to `elogit_types`, you will need to run the command

```
perl comserver.pl ELogitSrvr elogit_types.f90
```

again from the command prompt within the project directory. However, you will not be prompted again for the initial information shown above.

7.3.6   Building the Project in Visual Studio

The next step is to compile the COM server in Visual Studio. Follow these steps to create an out-of-process (`.exe`) COM server project.

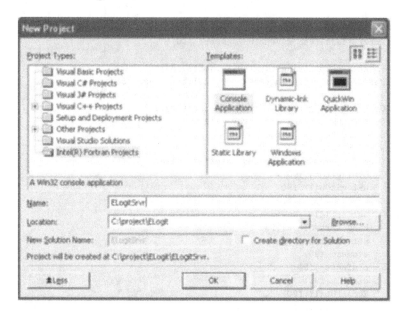

FIGURE 7.4. Selecting the project type in the Project Wizard.

1. Launch Visual Studio .NET 2003.

2. Press New Project.

3. In the resulting dialog, choose "Intel(R) Fortran Projects", and select "Console Application". Assign the project name ELogitSrvr and set the location to the previously defined project directory, ELogit (see Figure 7.4). Then press OK.

4. Select "Application Settings" and choose "Empty project" (see Figure 7.5). Then press Finish.

5. Add your existing code files to the project, including all of the automatically generated code.

   a. First, create a directory for automatically generated code. To do this, right-click on the project name in the Solution Explorer. Choose Add → NewFolder and name the folder "Autogen Files."

   b. Add the project source files to the project. To do this, right-click on the folder "Source Files," choose Add → Add existing item ... from the pop-up menu, and select the ELOGIT source files (you can select multiple files in the dialog by holding down the Ctrl key):

   constants.f90

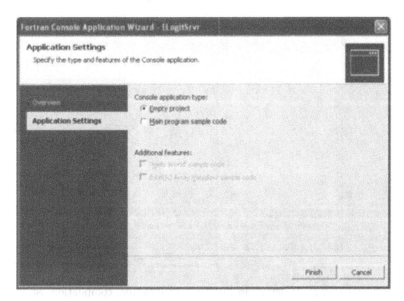

FIGURE 7.5. Selecting an empty project.

```
dynalloc.f90
elogit_gets.f90
elogit_io.f90
elogit_modelfit.f90
elogit_nullify.f90
elogit_puts.f90
elogit_types.f90
error_handler.f90
matrix.f90
variant_conv.f90
wls.f90
```

(This group includes the file `variant_conversion.f90`, which was automatically placed here when we ran the Perl script. The purpose of this file will be discussed later in Section 7.5.5.) Then press OK.

c. Remove the included source files from the compilation list. These files are part of the `elogit_types` module and have been effectively inserted into `elogit_types.f90` by `include` statements (Section 5.3.4); you need to remove them from the compilation list so that they will not be processed twice. To do this, go to the Solution Explorer and select the following files from the list:

```
elogit_gets.f90
elogit_io.f90
```

FIGURE 7.6. Removing included source files from compilation list.

```
elogit_modelfit.f90
elogit_nullify.f90
elogit_puts.f90
```

With all of these files selected, right-click on one of them and choose **Properties** from the popup menu. In the properties dialog, change the "Exclude File From Build" setting from **No** to **Yes** (Figure 7.6). Then press **OK**.

d. Add all of the automatically generated files to the project. To do this, right-click on the folder "Autogen Files," choose **Add →** **Add existing item ...**, and choose the automatically generated files (after selecting the "All files" filter in the dialog box):

```
clsfact.f90
clsfactty.f90
ELogitObjTY.f90
ELogitSrvrGlobal.f90
exemain.f90
IElogitObj.f90
server.idl
serverhelper.f90
UElogitObjTY.f90
UIElogitObj.f90
```

Then press **OK**.

FIGURE 7.7. Solution Explorer showing COM server project files.

e. Add the resource file to the project. Right-click on the folder "Resource Files," choose Add → Add existing item ..., and choose the file `server.rc`. Then press OK.

At this point, the Solution Explorer window should look like Figure 7.7. Notice that the files that we excluded from the compilation list have a different icon from the others.

6. Modify the project properties. First, open the project properties dialog. From the main menu, select Project → ELogitSrvr Properties. We want to change the properties for all configurations, so in the dialog change the Configuration: combo box to All Configurations. Then change the following properties:

a. Set up the application to use the Windows subsystem. Under the Linker item, select the System subitem. Change the SubSystem setting from Console to Windows as shown in Figure 7.8.

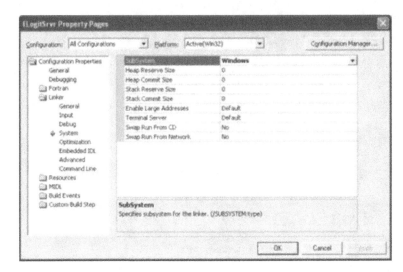

FIGURE 7.8. Setting linker properties.

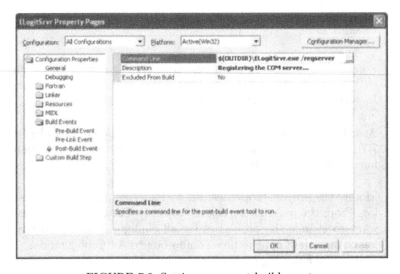

FIGURE 7.9. Setting up a post-build event.

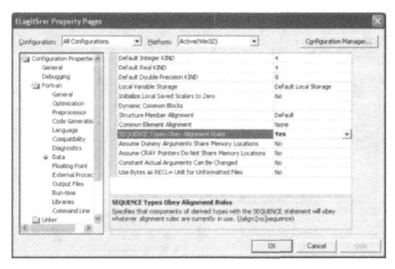

FIGURE 7.10. Setting Fortran data properties.

b. Set up a post-build event that registers the COM server. Under the Build Events item, choose the subitem Post-build Event. Under Command Line enter

```
$(OUTDIR)\ELogitSrvr.exe /regserver
```

and under Description enter

```
Registering the COM Server...
```

as in Figure 7.9.

c. Set the compiler to align **sequence** types properly. Under the Fortran item, select the Data subitem. In the category SEQUENCE Types Obey Alignment Rules, change the setting from No to Yes (Figure 7.10).

d. Set the resource compiler to look in the proper directory for the type library file. Under the Resources item, select the General subitem. In the category Additional Include Directories, add the text $(IntDir) (Figure 7.11).

After changing these settings, press OK to close the properties dialog. Finally, save the project settings by choosing from the menu File → Save All.

7.3.7 Building and registering the server

Once these steps have been taken, the project is ready to be built. From the menu, choose Build → Build ELogitSrvr. Output from the build should look like this:

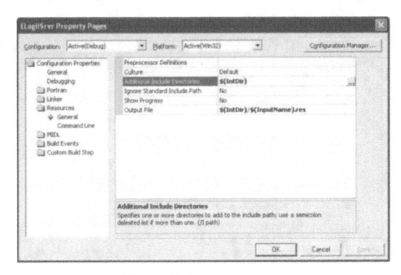

FIGURE 7.11. Setting resources properties.

```
---- Build started: Project: ELogitSrvr, Configuration: Debug Win32 ----

Creating Type Library...
server.idl
Compiling resources...
server.rc
Compiling...
error_handler.f90
constants.f90
clsfactty.f90
matrix.f90
wls.f90
dynalloc.f90
ELogitSrvrGlobal.f90
variant_conv.f90
elogit_types.f90
UELogitObjTY.f90
ELogitObjTY.f90
IELogitObj.f90
serverhelper.f90
UIELogitObj.f90
exemain.f90
clsfact.f90
Linking...
Registering the COM server....

Build log written to C:\project\ELOGIT\ELogitSrvr\Debug\BuildLog.txt
ELogitSrvr build succeeded.


--------------------- Done ----------------------

    Build: 1 succeeded, 0 failed, 0 skipped
```

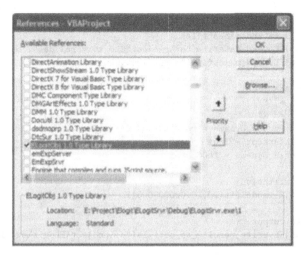

FIGURE 7.12. Adding a reference to the ELogit COM server in Excel VBA.

We do not need to manually register the new server because we added a post-build step for that purpose.

Once the Visual Studio project has been set up, it is reusable. The main Fortran module code and computational code can be edited. The automatically generated code files can be regenerated as needed by using the comserver.pl script. Then we simply need to rebuild the COM server by issuing the Build command.

7.3.8   Creating a Simple Client

To quickly test our new COM server, let us create a simple client in Excel that loads a small dataset from a file, specifies a model, fits the model, and writes the results to an output file. More sophisticated clients will be presented in the next chapter.

Open a new Excel spreadsheet and then open the Visual Basic editor window as described in Section 7.1.4. Add a reference to the COM server, which should be listed as "ELogitObj 1.0 Type Library" (Figure 7.12). Then add the following VBA code to Sheet1:

```
Sub elogit()

    ' Declarations
    Dim objEL As ELogitObj
    Dim resp(0 To 1) As Variant
    On Error GoTo errhandle

    ' Create object instance
    Set objEL = CreateObject("ELogitSrvr.ELogitObj")
```

```
          ' Read data and names files
          Call objEL.read_datafile("e:\project\ELogit\viral.dat", 5, 3, False)
          Call objEL.read_namesfile("e:\project\ELogit\viral.nam", 3)

          ' Load the model
          resp(0) = "Y"
          resp(1) = "N"
          objEL.response_byname = resp
          objEL.Intercept = True
          objEL.pred_byname = "LOG_DOSE"

          ' Run estimation
          Call objEL.modelfit(20, 0.0000000001)

          ' Write output
          Call objEL.write_results_to_outfile("e:\project\ELogit\excel.out")

          ' Exit the subroutine
          Exit Sub

errhandle:
          ' Handle errors
          If objEL.errMessagePresent Then
              MsgBox objEL.errMessage
          Else
              MsgBox Err.Description
          End If

End Sub
```

Let's examine this code in detail. The first dimensioning statement declares objEL to be a pointer to a future instance of an ELOGIT object. The second dimensioning statement declares resp to be a variant array of length 2. A variant is a very general data structure and will be explained in the next section. By convention, Visual Basic indexes elements of arrays beginning from 0, so the elements of this array are accessed in VBA as resp(0) and resp(1). The declaration

    On Error GoTo errhandle

instructs VBA to jump to the bottom portion of the script in the event of an error.

The first executable line creates an instance of the ELOGIT object and points objEL to it. Once this object exists, we can invoke its properties and methods. The read_datafile method is the COM server's version of the Fortran function read_elogit_datafile from Section 5.3.5; its arguments are the data file's name (given as a full path), the number of cases, the number of variables, and a logical value indicating whether case identifiers are present. You may recall that the Fortran function read_elogit_datafile had an additional dummy argument `err` for reporting error messages. Now that ELOGIT has become a COM server, the error messages are no longer an argument but a property of the ELOGIT object; they can be retrieved by invoking `objEL.errMessage` as shown near the bottom of this program. The call to the COM method `read_namesfile` follows a similar pattern; its arguments are the name of the variable names file (again as a full path) and the number of variables.

The next section of the VBA script loads the model specification. Here we see three examples of how properties are put into a COM object. Into the `response_byname` property, we put the `resp` array whose elements are "Y" and "N". Similarly, we put the logical value `True` into `Intercept` and the string "LOG_DOSE" into `pred_byname`.

After the model has been specified, we run the model fitting procedure by invoking `modelfit`. Because `modelfit` is a computational method rather than a property, it is preceded by `Call`. The arguments to `modelfit` specify the maximum number of iterations and the convergence criterion for the Newton-Raphson algorithm, as discussed in Section 5.5.4. A call to the `write_results_to_outfile` method reports the results to a file.

The `errhandle` section of this client swings into action if a call to any of the COM object's properties or methods produces an error. An error condition would be triggered, for example, if we had tried to invoke the `modelfit` method before the model was specified. In that case, the corresponding ELOGIT Fortran function described in Chapter 5 would store a message in the error handler and return a value of `RETURN_FAIL` rather than `RETURN_SUCCESS`. If a Fortran-generated error message has been stored, it will be retrieved by the VBA statement

```
MsgBox objEL.errMessage
```

and displayed in a message box. If no Fortran-generated message exists—which could happen, for example, if ELOGIT crashed unexpectedly due to an arithmetic exception—then the client will report whatever information is generated by VBA as it interacts with COM.

Notice that this simple client does not use any of the data-management capabilities of Excel. In fact, it doesn't really have anything to do with Excel. We are using Excel here merely as a vehicle for quickly writing and executing a VBA script that does the same thing as the ELOGIT console program; we could have done the same thing in SAS, MATLAB, S-PLUS, or many other environments. A more elaborate Excel client that inputs data from a spreadsheet and reports results to a spreadsheet will be developed in Chapter 8.

7.4 Exercises

1. What are the main differences between in-process and out-of-process COM servers? How is an in-process COM server different from a conventional DLL?

2. Describe the role of the interface in COM.

3. Define per-instance data and describe their role. What do you think the per-instance data are in the ELOGIT COM server? (Hint: Examine the Fortran code for the ELOGIT main program developed in Chapter 5, and try to identify which data structure or structures in that program have become the per-instance data.)

4. Suppose you have a COM server that defines an object class called Dog. An instance of this object in a VBA program could be referenced by a variable named objDog.

   a. If the Dog class has a method named bark, what VBA statement would you use to invoke this method on objDog?

   b. If there is a logical property called tailWagging, what VBA statement would you use to make objDog stop wagging its tail?

   c. If the Dog class has a method named rollover with an integer argument ntimes defining the number of times to roll over, how would you command objDog to roll over ten times?

5. Write an Excel macro that creates and uses two instances of the Magic 8-Ball object.

6. Microsoft Office has an animated Office Assistant that answers questions and provides suggestions. Depending on your settings, the Office Assistant may look like a paperclip, a bouncing red dot, etc. Some people find the Office Assistant annoying. Fortunately, you can hide it using the Help menu or by right-clicking on it. The Office Assistant is actually an object-oriented software component that can be controlled through COM. In Microsoft Excel, open a Visual Basic editor window. Open the Object Browser and look for the Assistant object, noting its properties and methods. Create a macro that turns on the assistant and causes it to perform various animations. (Hint: The Assistant is accessed through the application.assistant property.)

7. Experiment with the ELOGIT Excel client. See what happens when you try to use the COM server improperly (e.g., by trying to specify a model before data have been loaded).

8. Following the example of our simple Excel client, write another client that does the same thing in S-PLUS, R, MATLAB, or any environment of your choosing.

TABLE 7.1. Automatically generated code files.

Filename	Description
`clsfactty.f90`	The "class factory" module for creating object instances
`clsfact.f90`	Methods for `clsfactty.f90`
`exemain.f90`	(Out-of-process server only) Main entry point for the server's executable program
`dllmain.f90`	(In-process server only) Defines standard exported procedures for the server's DLL
`server.def`	(In-process server only) Declares exported procedures for the server's DLL
`server.idl`	Definition of the server's type library written in the IDL language
`serverhelper.f90`	Utility functions for the COM server
servername `Global.f90`	Module containing global data and related functions
classname `TY.f90`	Defines the per-instance data module
`I`*classname*`..f90`	Defines the interface module
`U`*classname*`TY.f90`	Module where the per-instance data are defined
`UI`*classname*`.f90`	Editable part of the interface containing functions that correspond to methods and properties

7.5 How the Fortran COM Server Works

7.5.1 Overview of the Automatically Generated Code

Using our Perl script, it is a simple matter to take a Fortran module written in our style and turn it into a COM server. We will now provide some insights into how the automatically generated code works. These insights will be helpful for creating COM servers and for writing COM clients.

The purpose of each of the automatically generated code files is summarized in Table 7.1. This table applies to ELOGIT and to any other project with appropriate strings substituted for *servername* and *classname*.

7.5.2   The Interface Definition Language File

Notice that some of the automatically generated files do not contain Fortran code. In particular, `server.idl` is written in a language called Interface Definition Language (IDL). IDL is crucial to COM because, as its name implies, it is used to define the COM interface. If you open this file with a text editor, you will see all of the puts, gets, and methods listed here along with their arguments.

There are three places in `server.idl` where the statement `uuid` appears followed by a long hexadecimal value in parentheses. They look like this:

```
uuid(42A7CC74-3E38-4BE4-AE15-EA5E92AE0BC1)
```

These are the universally unique identifiers (UUIDs) that allow a COM client to access the correct COM server. There is one UUID for the server itself, one for the interface, and one for the object class. No two COM servers will ever have the same UUIDs; it simply cannot happen. The UUIDs are generated by our Perl script the first time it is run using a built-in function from the Windows operating system. These same UUIDs remain associated with the COM server if the script is run again.

During the build process, `server.idl` is compiled by Visual Studio using the Microsoft IDL compiler into a binary version called a type library. The type library is the internal documentation of a COM server. It allows a client to find out exactly what the classes, properties, and methods are. If a method has arguments, these are listed in the type library as well. The type library indicates the nature of each property and argument (real, integer, logical, etc.).

The type library must be embedded inside the `.dll` or `.exe` file of a COM server. You may recall that when we built the ELOGIT COM server in Visual Studio, we had to add the resource file `server.rc` to the project. This file, which is placed in the project folder by our Perl script, directs Visual Studio to embed the compiled version of the type library into the COM server.

Extensive documentation about the IDL language, UUIDs, and type libraries is available online from the Microsoft Developer Network (MSDN) Library.

7.5.3   The Instance Code

The last four files listed in Table 7.1 pertain to the classes defined in the COM server. Like many COM servers, our ELOGIT server has only one class (`ELogitObj`). If more classes were present, these four files would be repeated for each additional class. Note the 'U' in the names of the last two files. These names, which are held over from Compaq's COM Server Wizard, were intended to signify that these files should be edited by the user (i.e., by you, the programmer) to customize the code within them.

For our ELOGIT project, no editing of these files was required. In fact, if you use our Perl script and follow the programming style we developed in Chapter 5, you may never need to edit the U-files. The contents of these files are helpful for understanding how Fortran interacts with COM, however, so let's open them and look inside.

The code inside UELogitObjTY.f90, which we call the instance code, creates a Fortran module named ELogitObj_USE. The preamble of this module has a number of use statements:

```
——————————————————————— UELogitObjTY.f90 ———————————

    use program_constants
    use error_handler
    use elogit_types
    use variant_conversion
```

If your COM server needs Fortran modules, this preamble is where they will be listed. Access to elogit_types is necessary to declare per-instance data and invoke the ELOGIT properties and methods that we wrote in Fortran. The variant_conversion module contains procedures that we have written to manage the passing of data arrays between COM and Fortran. The variant_conversion module, which will be explained later, is automatically added to every COM server project by the Perl script.

Below the use statements, the ELogitObj_USE module defines a single Fortran derived type, which becomes the per-instance data for the COM server. It also contains two Fortran procedures: a constructor, which determines the initial state of each new instance of the type, and a destructor, which is invoked whenever an instance is destroyed.

What exactly are ELOGIT's per-instance data? When we started to design the elogit_types module back in Section 5.3, we envisioned an object called an "ELOGIT session" that would store the dataset, the model specification, the results from the fitting procedure, and so on. We implemented this object in Fortran as a derived type called elogit_session_type with other derived types nested within it. The per-instance data for ELOGIT consist of one instance of the elogit_session_type and two instances of error_type (one to hold error messages, and the other to hold warning messages). The ELOGIT session, error messages, and warning messages are glued together to become the per-instance data for the ELOGIT object. This is how it looks inside UELogitObjTY.f90:

```
——————————————————————— UELogitObjTY.f90 ———————————

    type ELogitObj_InstanceData
        sequence
        type(session_type) :: session
        type(error_type) :: err
        type(error_type) :: warn
    end type
```

Looking below the `contains` statement, we see two Fortran procedures. The first, `ELogitObj_CONSTRUCTOR`, is a function that runs automatically whenever an ELOGIT object is instantiated. It has a single dummy argument of type `ELogitObj_InstanceData`. The code inside this function, which was automatically generated by the Perl script, resets the error and warning messages and nullifies the ELOGIT session by a call to the `nullify_elogit_session` function that we wrote in Chapter 5.

```
─────────────────── UELogitObjTY.f90 ───────────────────
function ELogitObj_CONSTRUCTOR (ObjectData) result (hresult)
  use ifwinty
  implicit none
  type(ELogitObj_InstanceData) ObjectData
  !dec$ attributes reference :: ObjectData
  integer(LONG) hresult

  hresult = S_OK

  ! Add field initialization code - reset and nullify methods
  call err_reset(ObjectData%error)
  call err_reset(ObjectData%warn)

  if(nullify_elogit_session(ObjectData%session,ObjectData%error) &
     == RETURN_FAIL) then
     hresult = E_FAIL
  end if

end function
```

This function returns an integer value `S_OK` if it runs successfully or `E_FAIL` if something goes wrong. `S_OK` and `E_FAIL` are integer parameters defined in the Intel Visual Fortran module `ifwinty` and are also recognized by COM; we'll discuss these further in Section 7.5.6.

The Perl script has also written a destructor that runs whenever an instance of the ELOGIT object is destroyed. This destructor first wipes out any error or warning messages that may be present. It then calls the function `nullify_elogit_session`, which we wrote in Chapter 5, to wipe out any data stored in the session. Notice that this is a subroutine rather than a function; COM does not expect a destructor to return a value.

```
─────────────────── UELogitObjTY.f90 ───────────────────
subroutine ELogitObj_DESTRUCTOR( ObjectData )
   implicit none
   type(ELogitObj_InstanceData) ObjectData
   !dec$ attributes reference :: ObjectData
   !  field cleanup code

   integer(our_int) :: ret
   call err_reset(ObjectData%error)
   call err_reset(ObjectData%warn)
```

```
        ret=nullify_elogit_session(ObjectData%session,ObjectData%error)
end subroutine
```

7.5.4   The Interface Code

The other file whose name begins with U, UIELogitObj.f90, creates the
public interface for the ELOGIT object. (The I following the U stands
for "interface.") This is where all of the puts, gets, and other methods
for the ELOGIT object are defined. When we wrote the elogit_types
module back in Chapter 5, we made it object-oriented and carefully implemented all of the puts, gets, and methods as Fortran functions. This took
a bit of effort. Now this effort has really paid off because the functions in
UIELogitObj.f90 are nothing more than wrappers for the public functions
in elogit_types. The Perl script was able to generate all of this wrapper
code automatically.

Let's look at one of the functions in UIELogitObj.f90 to see what is
happening.

```
————————— UIELogitObj.f90 —————————
! IELogitObj_put_intercept interface function for put_elogit_intercept
! in module elogit_types
function IELogitObj_put_intercept(ObjectData, VALUE ) result (hresult)
    use ELogitObj_Types
    implicit none
    type(ELogitObj_InstanceData) ObjectData
    !dec$ attributes reference :: Objectdata
    LOGICAL(2) ,intent(in) :: VALUE
    integer(LONG) hresult

! method implementation

    logical :: bVal
    bVal = VALUE
    hresult=S_OK

    ! put data into server
    if(put_elogit_intercept(bVal,ObjectData%session,ObjectData%error)&
        == RETURN_FAIL) then
        hresult=E_FAIL
    end if

end function
```

This code simply takes a logical value supplied by a COM client and
passes it along to our own function put_elogit_intercept. A minor com-
plication is that the local variable bVal is needed to convert the value to the
default logical kind expected by put_elogit_intercept. Now let's look at
the get method for this same property.

```
───────────────────────── UIELogitObj.f90 ─────────────────────────
! IELogitObj_get_intercept interface function for get_elogit_intercept
! in module elogit_types
function IELogitObj_get_intercept(ObjectData, VALUE ) result (hresult)
    use ELogitObj_Types
    implicit none
    type(ELogitObj_InstanceData) ObjectData
    !dec$ attributes reference :: Objectdata
    LOGICAL(2) ,intent(out) :: VALUE
    integer(LONG) hresult

 ! method implementation

    logical :: bVal
    hresult=S_OK

    ! retrieve data from server
    if(get_elogit_intercept(bVal,ObjectData%session,ObjectData%error)&
        == RETURN_FAIL) then
      hresult=E_FAIL
      VALUE = .false.
    else
      VALUE = bVal
    end if

end function
```

This is the same idea, only in reverse. Notice how this function handles errors. If everything works normally, the returned value is the integer parameter S_OK. If an error arises—for example, because the client tries to get the intercept property before any data are loaded or before a response variable has been specified—then the returned value is E_FAIL.

A common feature of all these puts and gets is that each function has two dummy arguments. The first argument is the per-instance data, and the second is of whatever type is needed for that property. These two arguments are passed to Fortran whenever a COM client invokes a property using the dot notation, as in objEL.Intercept.

Finally, let's look at one of the computational methods.

```
───────────────────────── UIELogitObj.f90 ─────────────────────────
! IELogitObj_modelfit interface function for run_elogit_modelfit
! in module elogit_types
function IELogitObj_modelfit(ObjectData, maxits, eps) result (hresult)
    use ELogitObj_Types
    implicit none
    type(ELogitObj_InstanceData) ObjectData
    !dec$ attributes reference :: Objectdata
    INTEGER(4), intent(in), optional :: maxits
    REAL(8), intent(in), optional :: eps
    integer(LONG) hresult
    hresult=S_OK
```

```
! method implementation

  if(run_elogit_modelfit(ObjectData%session,ObjectData%error,&
     ObjectData%warn, maxits=maxits, eps=eps)==RETURN_FAIL) then
     hresult = E_FAIL
  end if

end function
```

This method simply invokes the function **run_elogit_modelfit**, passing to it the **session**, **error**, and **warn** components of the per-instance data. Upon completion, this function returns an integer value of **S_OK** or **E_FAIL**. Notice that the optional argument **maxits** is a long (32-bit) integer and **eps** is a double-precision (64-bit) real. Those kinds of data are easily recognized and handled by COM. Some of the properties and arguments in the file **UIELogitObj.f90** are arrays, however; those are passed as variants, as we will describe momentarily.

The properties of the ELOGIT object are summarized in Table 7.2. All of the other methods are listed in Table 7.3. Whenever you build a COM server, we suggest that you generate tables like these. If the user of your server knows about COM and has a basic understanding of what the server does, he or she will need very little documentation beyond these tables to figure out how to use it.

7.5.5   Passing Arrays as Variants

To help ensure that data arrays are passed properly between Fortran and the COM client, we have decided to package them as variants. Variants are a very general type of data structure whose format is part of the COM standard. Variants are self-defined; they carry all the information necessary to identify the types and dimensions of the data within. Variants are not a part of the standard Fortran language, but the Intel Visual Fortran COM library **IFCOM** has implemented them as a Fortran derived type called **VARIANT**.

One of the components of Intel's **VARIANT** type, **VT**, indicates what kind of data are stored in the variant. The types most useful for our purposes are shown in Table 7.4. If a variant **myVariant** is a scalar integer, for example, then **myVariant%VT** will have the value 3. If the variant is an array, then the **VT_ARRAY** must be added to another parameter to identify it. For example, the value of **VT** for an array of double-precision reals is **VT_ARRAY + VT_R8** = 8192 + 5 = 8197, or 2005 in hexadecimal format.

The other important component of the variant type is **VU**, which contains the actual value of the variant. **VU** is a special type called a *union*, a single variable that can represent multiple derived types or facets. One facet is

TABLE 7.2. ELOGIT COM server properties.

Property	Description	Fortran type	Access*
data_matrix	Rectangular dataset	type(VARIANT)	RW
caseid	Case identifiers	type(VARIANT)	RW
var_names	Variable names	type(VARIANT)	RW
response_bycol	Response, column no.	type(VARIANT)	RW
response_byname	Response, var. name	type(VARIANT)	RW
intercept	True if intercept	logical(2)	RW
pred_bycol	Predictors, column no.	type(VARIANT)	RW
pred_byname	Predictors, var. name	type(VARIANT)	RW
beta	Estimated coefficients	type(VARIANT)	RW
ncase	No. cases in dataset	integer(4)	R
nvar	No. variables	integer(4)	R
beta_names	Names for coefficients	type(VARIANT)	R
iter	No. iterations	integer(4)	R
converged	True if converged	logical(2)	R
cov_beta	Covariance matrix	type(VARIANT)	R
loglik	Loglikelihood	real(8)	R
X2	Pearson statistic	real(8)	R
G2	Deviance statistic	real(8)	R
df	Degrees of freedom	real(8)	R
errMessage	Error message	character(*)	R
errMessagePresent	True if message present	logical(2)	R
warnMessage	Warning message	character(*)	R
warnMessagePresent	True if message present	logical(2)	R

* RW = read/write; R = read-only.

for 32-bit integers, one facet is for double-precision reals, and so on. The facets of VU are shown in Table 7.5.

To see how VT and VU work together, suppose that you want to store the square root of 2 in double precision as a variant. This is how you would do it in Intel Visual Fortran:

```
! use the IFCOM library
use IFCOM

! declare a variant
type(VARIANT) :: myVariant

! specify that an 8-bit real (double precision)
! number is to be stored in it
myVariant%VT = VT_R8

! store the number
myVariant%VU%DOUBLE_VAL = sqrt(2.D0)
```

TABLE 7.3. ELOGIT COM server methods.

Method / Description	Arguments	Type
modelfit		
Fits logistic regression model by ML	maxits	integer(4)
	eps	real(8)
reset		
Nullifies per-instance data		
read_datafile		
Reads data matrix from external ASCII file	data_file_name	character(*)
	nkase	integer(4)
	nvar	integer(4)
	case_id_present	logical(2)
read_namesfile		
Reads variable names from ASCII file	names_file_name	character(*)
	nvar	integer(4)
write_results_to_outfile		
Writes summary of model fit to ASCII file	output_file_name	character(*)

TABLE 7.4. Intel Visual Fortran's variant types.

Parameter	Actual value	Description
VT_I4	3	long integer
VT_R4	4	real
VT_R8	5	double precision
VT_BSTR	8	character string
VT_BOOL	11	logical
VT_VARIANT	12	VARIANT
VT_ARRAY	8192 (hex: 2000)	array

TABLE 7.5. Facets of a variant VU component.

Component	Description
VU%LONG_VAL	long integer value
VU%CHAR_VAL	character value
VU%SHORT_VAL	short integer value
VU%FLOAT_VAL	real value
VU%DOUBLE_VAL	double-precision value
VU%BOOL_VAL	logical (boolean) value
VU%SCODE_VAL	long integer value
VU%DATE_VAL	numeric time and date value
VU%PTR_VAL	pointer value

Variant Arrays

There are many ways to package arrays as variants. In our experience, applications that support COM tend to conform to the conventions of VBA as implemented in Microsoft Excel because many software vendors want to make their applications compatible with Excel. In fact, they often illustrate their COM automation capabilities with examples involving Excel. When an Excel spreadsheet passes data from a range of cells to a COM server, it packages them as *a variant containing a two-dimensional array of variants*. Therefore, we will also follow this convention. If a client has a different method for packaging arrays as variants, it is often possible to coerce those variants into the Excel form using a function called `variantChangeType` provided by the COM infrastructure of Windows.

If this is starting to sound too complicated, don't worry; we have written a Fortran module called `variant_conversion` to deal with variant arrays. The source code for this module is available from our Web site, and our Perl script automatically adds this module to every COM server project. The module has two public functions, and both are generic. The first function, which extracts a Fortran array from a variant of the type used by Excel, begins like this:

```
integer(our_int) function get_variant_array( varArray, &
    pArr, err )
```

The other function converts a Fortran array into an Excel-compliant variant:

```
integer(our_int) function put_variant_array( varArray, &
    pArr, err )
```

In each case, the first dummy argument, `varArray`, is an Intel Visual Fortran `VARIANT`. The last argument, `err`, is an `error_type` from our error-handling module. The other argument, `pArr`, is a pointer to a rank-one or rank-two array of type `integer(our_int)`, `real(our_dble)`, or `character(len=*)`. The array may be indexed beginning from 0 or 1.

Converting Variants to Fortran Arrays

Now let's look at some interface code that converts variants to arrays. The put method for the `data_matrix` property is shown below.

```
───────────── UIELogitObj.f90 ─────────────
! IELogitObj_put_data_matrix interface function for
! put_elogit_data_matrix in module elogit_types
function IELogitObj_put_data_matrix(ObjectData, VALUE ) result (hresult)
    use ELogitObj_Types
    implicit none
    type(ELogitObj_InstanceData) ObjectData
    !dec$ attributes reference :: Objectdata
    type(VARIANT) ,intent(in) :: VALUE
```

```
integer(LONG) hresult
real(our_dble), pointer :: pArr(:,:)

nullify(pArr)
hresult=S_OK
if(get_variant_array(VALUE, pArr, ObjectData%error) &
    == RETURN_FAIL) then
  hresult = E_FAIL
else if(put_elogit_data_matrix(pArr, ObjectData%dataset, &
    ObjectData%error) == RETURN_FAIL) then
  hresult=E_FAIL
end if
if (associated(pArr)) deallocate(pArr)

end function
```

This function accepts a data matrix that was sent by the COM client as
a variant, converts it to a Fortran array, and loads it into the per-instance
data using the `put_elogit_data_matrix` function. The corresponding get
function does the same thing in reverse, calling `get_elogit_data_matrix`
followed by `put_variant_array`.

7.5.6   How the COM Server Handles Errors

The COM standard dictates that any call to a property or method must
return a value called an `hresult`. An `hresult` is a 32-bit integer that
functions as an error code. Many of these codes have been declared as
named constants by COM and by Visual Fortran's COM libraries. We will
use just two of these values: `S_OK`, which indicates a successful completion,
and `E_FAIL`, which indicates a fatal error.

The `hresult` returned by a property or method enables the COM client
to handle errors. You may not actually see the `hresult` in the COM client's
code, but it's there. For example, suppose that our Excel client for ELOGIT
had tried to invoke the `modelfit` method before the model had been specified. The COM interface would have then returned an `hresult` of `E_FAIL`.
Excel would have automatically detected the error and taken action because of our declaration

```
On Error GoTo errhandle
```

near the beginning of the macro.

The `hresult` tells the client if an error has occurred, but it provides no
further details. Fortunately, the Fortran module for handling errors that we
have used throughout this book is an excellent tool for sending a text message back to the client. Our Perl script, which assumes that you have used
`error_handler`, automatically includes an `error_type` variable in the per-
instance data and creates two read-only properties: `errMessagePresent`

and `errMessage`. Invoking `errMessagePresent` will obtain a logical value indicating whether any messages have been stored, and `errMessage` obtains a stored message as a text string. Here is the interface code that gets `errMessage`:

```
──────────────── UIELogitObj.f90 ────────────────
! IELogitObj_get_errMessage
function IELogitObj_get_errMessage (ObjectData, Value) result (hresult)
    use ELogitObj_Types
    implicit none
    type(ELogitObj_InstanceData) ObjectData
    !dec$ attributes reference :: Objectdata
    CHARACTER(*), intent(out) :: VALUE
    integer(LONG) hresult
    call err_get_msgs(ObjectData%error, VALUE, "PC")
    call err_reset(ObjectData%error)
    hresult = S_OK
end function
```

This function calls the subroutine `err_get_msgs` in the `error_handler` module to retrieve all the lines of text currently stored in the error handler object. These lines are returned as a single character string with embedded hard carriage returns so that the lines will be displayed properly in a Windows environment. Notice that this function also resets the error handler, wiping out the messages that have just been reported.

Our per-instance data for ELOGIT also contain a second instance of the `error_type` to hold warning messages generated by nonfatal events. When the `read_namesfile` method is invoked, for example, a warning is generated if some variable names were too long and had to be truncated. If you are creating a COM server and any of your public procedures has as a dummy argument a second `error_type` called `warn`, the Perl script will automatically create two additional read-only properties, `warnMessagePresent` and `warnMessage`, so that a client can read the warnings.

7.6 Distributing and Installing COM Servers

After creating, debugging, and testing a COM server, it is ready to be distributed to interested users. When a COM server is to be installed on another computer, there are several things to keep in mind.

First, the COM server should be built in "Release" mode. A COM server (or any other executable or DLL) built in "Debug" mode requires special libraries that ordinary users will not have. Those libraries cannot be freely distributed. Executables built in "Release" mode also tend to be smaller and run faster because the internal code has been optimized.

Second, care must be taken to handle registration. As we mentioned previously, a COM server must be registered before it can be used on a computer. This is done by invoking the server at the command line with the /regserver option:

```
myserver /regserver
```

Before you remove a COM server from a computer, we strongly recommend that you unregister it, like this:

```
myserver /unregserver
```

Registration places entries into the Windows registry, and unregistration removes those entries. The entries include the UUIDs for the server, interface, and class.

Failure to unregister a server may cause problems. Suppose that a reader of this book downloads ELogitSrvr.exe from our Web site and registers it on his or her computer. Later, that same user downloads the source code for the ELogitSrvr project, runs the Perl script for a new project, and builds the ELogit COM server. What happens? The Perl script generates a server with a completely different set of UUIDs. Now there are two COM servers with the name ELogitSrvr.ELogitObj that are separately registered. When the user runs one of the client programs, there is some ambiguity as to which COM server is actually being used by the client. Let us then suppose that the reader deletes the first server without unregistering it. The registry still believes the server is there, but it is not. The client may no longer work if it is attempting to access the nonexistent server. The only way to fix such a problem is to manually edit the registry (which the authors do not recommend) or to redownload the original COM server to the same location as before, unregister it, and then delete it.

As an interesting but potentially important side note, two servers with the same name can be registered if they are given a different version number. For example, the reader mentioned above could give the version number as "2.0" when prompted by the Perl script. This version of ELogitSrvr will have an identity unique from Version 1.0, the downloaded server. Clients would then need to distinguish among these versions. For example, the Excel client code could issue the command

```
Set objEL = CreateObject("ELogitSrvr.ELogitObj.2.0")
```

to access Version 2.0. In this case, the new version becomes the default, so the .2.0 designation is optional. However, to access the older version, it is now necessary to issue the command

```
Set objEL = CreateObject("ELogitSrvr.ELogitObj.1.0")
```

because it is no longer the default version. Client code for other applications would also need to be changed accordingly.

Another potential problem arises from renaming or moving the COM server to a different file folder. Before doing so, it is very important to unregister the COM server. Then, after renaming or moving the file, it must be reregistered. To summarize, whenever a COM server is to be moved, deleted, or renamed, it must first be unregistered.

Distributing COM servers can be greatly aided by the use of an installation tool, especially when it is accompanied by other programs, documents, and other items. One of the simplest and cheapest tools is the self-extracting "zip" file. Installation tools can be configured to automatically register COM servers on the target computer. The more sophisticated ones also allow the user to uninstall your software, which will then automatically unregister any COM servers that were distributed with it.

7.7 Additional Exercises

1. Explain why COM servers are typically designed for situations where property and method calls are relatively infrequent. If you want to create a library of fast computational procedures that may be called many times in rapid succession, what should you do?

2. In the Magic8 COM server, create properties that enable a client to cheat. For example, implement a read-write property `valence` that takes integer values -1, 0, or 1 to indicate the client's desire for an answer that is negative, vague, or positive. Then implement a logical read-write property `cheat` that, if set to `True`, instructs the server to require future messages to have the desired valence.

3. What is the purpose of an IDL file? Open the IDL file from the ELOGIT COM server, examine it, and try to figure out what it does.

4. In the ELOGIT interface code, the `IELogitObj_get_errMessage` function has a dummy character-string argument of assumed length. Our Excel VBA client that invokes the `errMessage` property does not specify a length for the message string either. This suggests that some additional code must exist between the VBA client and `IELogitObj_get_errMessage`, which, among other things, determines a maximum length for the message string. Locate this code in the automatically generated files and try to figure out what it does.

5. Create a COM server that performs simple linear regression on data pairs (x_i, y_i). $i = 1, \ldots, n$. Follow these steps.

   a. Write a Fortran computational procedure that accepts as input two arrays $(x_1, \ldots, x_n)$ and $(y_1, \ldots, y_n)$ and computes the estimated intercept and slope.

TABLE 7.6. Data for testing the regression COM server.

x	y
2.42	0.52
0.61	0.97
0.18	0.61
6.49	2.60
1.92	0.96
0.69	0.32
4.90	2.35
3.30	1.39
4.44	2.20
0.16	0.67
2.02	1.69
4.67	2.46
0.77	1.29
7.11	3.32
2.52	1.77
3.47	2.05

b. Create an object-oriented Fortran module for simple linear regression that uses your procedure. Implement read-write properties x and y, a computational method lsfit, and read-only properties intercept and slope.

c. Turn your module into an out-of-process COM server, using our Perl script to generate the required code.

d. Create a test client in Excel that performs linear regression on the $n = 16$ pairs shown in Table 7.6. Verify that the estimated intercept and slope are 0.5418 and 0.3613, respectively.

6. In Exercise 6 from Section 5.6, you were asked to develop an object-oriented console program for performing the chi-square test for independence on a two-way table. If you have done this, turn your console program into a COM server. Test the server by creating a simple client in Excel.

8
Creating COM Clients

As we have seen, creating a Fortran COM server takes a fair amount of effort. You must organize your data into objects and properties and create an interface with puts, gets, and methods. The build process for a COM server is more complicated than for a console program or conventional DLL. Once the COM server exists, however, you begin to reap the fruits of your labor. COM clients are much easier to create than servers, and they can be written in a surprisingly large number of environments.

In this chapter, we show by example how to create COM clients in Excel, S-PLUS, R, SAS, SPSS, and MATLAB. With well-written clients, users of those packages may call your statistical routines without needing to know anything about COM. We conclude this chapter by showing how to create a complete, stand-alone statistical application for Windows with a graphical user interface (GUI) using Visual Basic .NET®.

8.1 An Improved Client for Excel

8.1.1 Excel As a Graphical User Interface

In the last chapter, we created a bare-bones client in Excel for testing the ELOGIT COM server. Our client's interaction with Excel was minimal; it read data from files and wrote results to a file. Most users of Excel would prefer to enter their data and specify the modeling options directly on a spreadsheet. After fitting the model, the results can be reported back to

FIGURE 8.1. ELOGIT Excel client spreadsheet.

the spreadsheet where they can be easily examined and turned into graphs, tables, and so on.

Most GUIs for Windows applications have been written in languages such as Visual Basic and Visual C++. If you are not familiar with any of those languages, writing your own GUI can be a daunting task. The material at the end of this chapter will help you get started. But if you want to create a simple GUI as quickly as possible, perhaps you should consider doing it in Excel.

8.1.2   Starting to Write the Client

Let's begin to write a more elaborate ELOGIT client that uses the graphical features of Excel. As a first step, open Excel with a fresh, new workbook. In the Sheet1 spreadsheet, enter the viral assay experiment data from Table 5.1 in the range of cells from B2 to D7. Enter names for the variables in cells B1 to D1. Then fill in the rest of the information shown in Figure 8.1 (ignoring the buttons for now).

We are almost ready to add the VBA code behind this client application. Before we start, however, let's do something to make that code simpler and easier to read: apply names to ranges of cells. In the spreadsheet, select the cells that contain the data values (B2–D7) and type **DataMatrix** in the Name Box as shown in Figure 8.2. This sets up a one-to-one correspondence

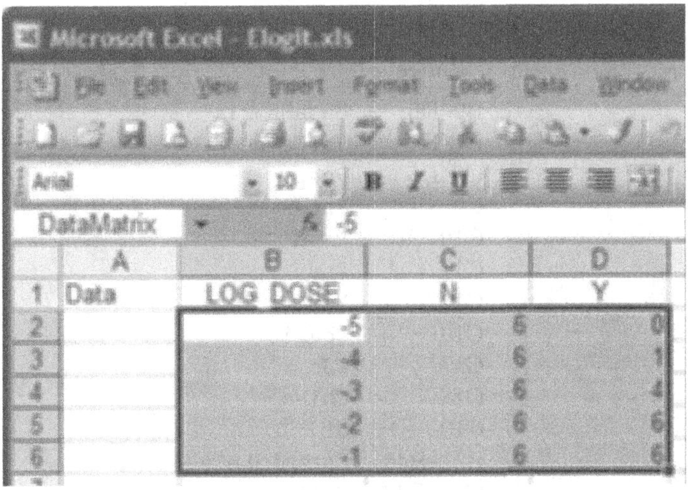

FIGURE 8.2. Naming a range in Excel.

in Excel between the range name **DataMatrix** and the cell range B2 to D7. (Another way of defining or redefining named ranges is to select a range of cells and then choose the menu items Insert → Name → Define... .) In the same way, apply descriptive names to the rest of the cell ranges listed in Table 8.1.

Once the names have been defined, go to the Excel menu and choose Tools → Macro... → Visual Basic Editor. In the Visual Basic editor window, select Sheet1 to view the macro (VBA code) page associated with the workbook sheet Sheet1. Then add a reference to the ELOGIT object as follows. In the VBA editor, choose Tools → References... from the menu, check the box next to "ELogitObj 1.0 Type Library" in the selection box, and press OK. The VBA code will now have access to the ELOGIT COM server.

Returning to the VBA editor, enter the lines of code shown below.

———————— elogit.vba ————————

```
Sub runELogit()

    ' Declare and instantiate an ELogit object
    Dim objEL As ELogitObj
    Set objEL = CreateObject("ELogitSrvr.ELogitObj")

    ' Load the data matrix
    objEL.data_matrix = Range("DataMatrix").Value

    ' Load variable names
    objEL.var_names = Range("VarNames").Value

    ' Specify response variable
    objEL.response_byname = Range("Resp").Value
```

TABLE 8.1. Names for cell ranges.

Range	Name
B2–D6	DataMatrix
B1–D1	VarNames
C8–D8	Resp
C9	Inter
C10	Pred
C11	Eps
C12	Maxits
C16	Loglik
C17	GSq
C18	ChiSq
C19	df
B21–B46	EstimLabels
C21–C46	Beta
D21–D46	StdErr
E21–E46	Ratio

```
    ' Specify intercept
    objEL.Intercept = Range("Inter").Value

    ' Specify predictors
    objEL.pred_byname = Range("Pred").Value

    ' Fit model
    Call objEL.modelfit(Range("Maxits").Value, _
        Range("Eps").Value)

    ' Display status
    If objEL.converged Then
        Call MsgBox("Algorithm converged in " & _
            objEL.iter & " iterations.")
    Else
        Call MsgBox("Algorithm did not converge by " & _
            objEL.iter & " iterations.")
    End If

    ' Clear the object
    Set objEL = Nothing

End Sub
```

You will notice that whenever you type objEL., the editor automatically displays the properties and methods of the ELogitObj object. You can either select from the menu by pointing and clicking or you can start typing a method property name after objEL. and the editor will select it for you. Once the selection has been made, you can press space or Enter and the editor will complete the method or property name for you. So, for example, instead of typing objEL.dataMatrix, you can simply type objEL.d followed by space or Enter. This feature of the VBA editor is designed to help you write code faster and with fewer errors.

We are now ready to run this VBA client for the first time. Place the cursor on any line inside the subroutine and press the run button ▸ or the F5 keyboard key. What happens? If no errors are encountered, a dialog box appears to inform you that the model-fitting algorithm converged in eight iterations. If your code contains errors, an error message box will appear to help you correct the problem.

8.1.3   How Did It Work?

This is what happens when you tell Excel to run the VBA program. In response to the first executable line,

```
Set objEL = CreateObject("ELogitSrvr.ELogitObj")
```

Excel asks the Windows operating system to launch the ELOGIT COM server (ELogitSrvr.exe) and create an instance of the ELogitObj class. When the new instance is created, the COM server immediately runs the constructor function associated with the ELogitObj class. The COM server then returns a pointer—in this case, an integer value indicating the COM server's memory address—and stores it in the VBA variable objEL.

The next few lines invoke properties of objEL. For example,

```
objEL.data_matrix = Range("DataMatrix").Value
```

puts data into data_matrix. The right-hand side of this statement invokes the spreadsheet's Range property, with the desired range's name as its argument. This property returns a Range object that has the property Value. The whole expression on the right-hand side therefore evaluates to the set of data values currently stored in cells B2–D6. When setting the property data_matrix, the ELOGIT COM server expects to receive data of type(VARIANT) as explained in Section 7.5.5. When Excel passes the contents of B2–D6, the COM infrastructure is able to coerce these data into the expected type, provided that the data in B2–D6 are numeric. If any of the cells contains non-numeric data (e.g., text), Excel will return an error.

The line below instructs the ELOGIT COM server to fit the model.

```
' Fit model
Call objEL.modelfit(Range("Eps").Value, _
     Range("Maxits").Value)
```

The underscore (_) is VBA's continuation symbol. Notice again how we have used Range and Value to specify arguments for the modelfit method. In this case, the COM server is expecting a double-precision real number and a long integer, respectively. When Excel passes the contents of the cell ranges Eps and Maxits, COM is smart enough to coerce the numeric data in those cells into the proper types. As long as the data can be made compatible with what ELOGIT is expecting, COM will make it work.

This next portion of code reports information on convergence.

```
' Display status
If objEL.converged Then
    Call MsgBox("Algorithm converged in " _
    & objEL.iter & " iterations.")
Else
    Call MsgBox("Algorithm did not converge by " _
    & objEL.iter & " iterations.")
End If
```

Notice how the object-oriented nature of COM servers makes the code very neat and readable. The converged property returns a boolean value (True or False), and the iter property returns an integer.

The last executable statement releases the ELOGIT object.

```
Set objEL = Nothing
```

Setting an object to Nothing removes the reference to the object so that the COM server can free its allocated memory. A COM server keeps track of how many instances of its objects are active at any time; if that number drops to zero, the server automatically shuts itself down.

8.1.4 Debugging the Client and Server

The Visual Basic editor in Excel also functions as a debugger. It provides standard debugging features, including breakpoints, stepping, and variable watching. These are invoked through the keyboard's function keys. For example, you can step through the client code one line at a time by pressing F8.

When developing COM clients, it may be helpful to debug the client and server simultaneously. While invoking properties and methods, you can directly observe how the client passes data to the server. To see how this works, leave your Excel and Visual Basic windows open, and open the ELogitSrvr project (as created in Chapter 7) in Visual Studio .NET. From the Visual Studio menu, choose Build → Configuration Manager... and, in the dialog window, choose Debug from the list in the "Active Solution Configuration" combo box. Next, set some breakpoints in the source code. Open the file UIELogitObjTY.f90, place your cursor on the line

```
hresult = S_OK
```

in the constructor function, and press the F9 key. A red circle will appear to the left of the code line (Figure 8.3). Place another breakpoint in the

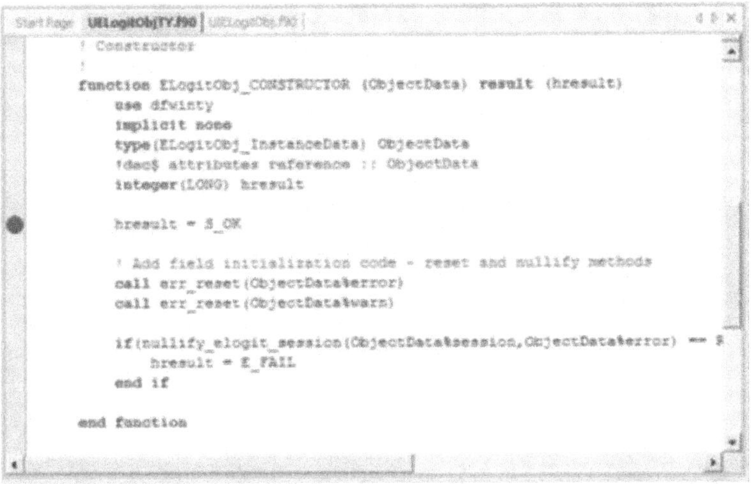

FIGURE 8.3. Setting a breakpoint in the constructor.

file `UIElogit.f90` in the line

```
hresult = S_OK
```

within the function `IELogit_put_data_matrix`. Then launch the COM server by pressing the F5 key or the Run button in Visual Studio. What happens? The COM server is launched and loaded into memory, but no instances of its class are yet created. The COM server sits in memory, waiting for a client application to use it.

Now go back to the Excel VBA window. Press F5 to start the script. Immediately, you notice that the Visual Studio window has paused at the breakpoint you set in the `ELogitObj_CONSTRUCTOR` function. The COM server invoked this function as the new instance of the `ELogitObj` class was created.

In Visual Studio, press F5 again to continue. What has happened now? The Excel macro (VBA code) has proceeded to invoke the `data_matrix` property. In Visual Studio, the debugger is now paused at the breakpoint you set in `IELogit_put_data_matrix`. Let's execute the statements in this function one at a time. Press F10, which instructs the debugger to proceed to the next code line. That line, which calls the `get_variant_array` function from the module `variant_conversion`, converts the Excel-supplied `type(VARIANT)` into a Fortran array. It takes `pArr` (an unallocated pointer to a rank-two double-precision array) as an argument, allocates it to the correct size, and loads the data values into it.

We can now see whether Excel is passing the data matrix correctly. Press the F10 key again to complete execution of the current line. In the bottom left pane of the Visual Studio window, select the Locals tab. Then expand

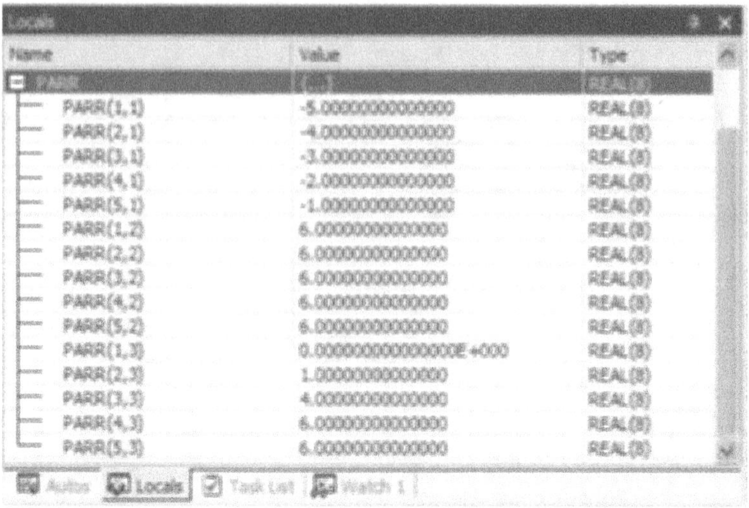

FIGURE 8.4. Contents of the data array after variant conversion.

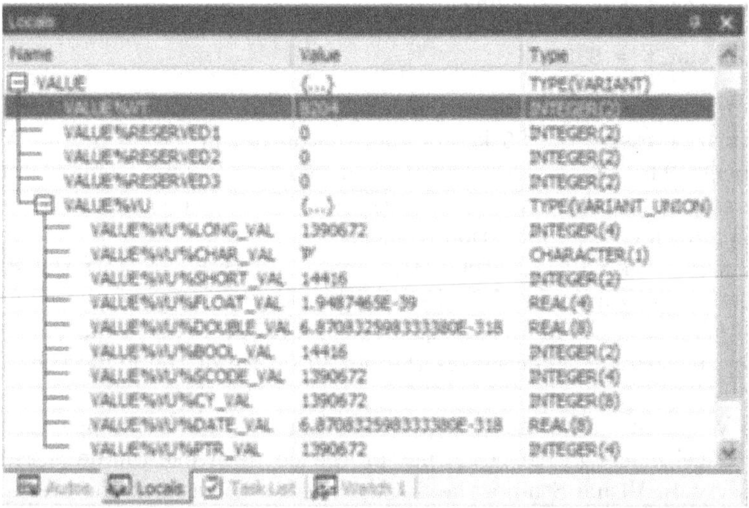

FIGURE 8.5. Inside the type(variant) containing the data matrix.

the PARR item to see that the contents of this Fortran array match the data in the spreadsheet (Figure 8.4).

While we're in the debugger, let's also peek inside the variable VALUE, the data of type(VARIANT) passed from Excel. It should appear as shown in Figure 8.5. The first component of this variant is VALUE%VT. This is the identifier that signals what the variant contains. In this case, the value of

FIGURE 8.6. Contents of data_matrix in the per-instance data.

VALUE%VT is 8204, which is the sum of the array constant (VT_ARRAY=8192) and the variant constant (VT_VARIANT=12). The contents of the variant are listed under VALUE%VU. Here we see the various possible incarnations of the variant as a long integer, character, short integer, and so on. Because this is an array, the one we are interested in is VALUE%VU%PTR_VAL. This variant is actually a pointer containing the memory address of the array. Our Fortran function get_var_array uses this address to obtain the contents of the variant array. Different client applications package these arrays differently, and get_var_array is designed to deal with all the various kinds. Excel, for example, always passes the contents of a range of cells as a rank-two array, even if the dimensions of the range are 1×5 or even 1×1. (If you are curious to see how get_var_array works, you may step through its operation in the debugger by pressing the F11 key instead of F10. This tells the debugger to step into a called procedure's code rather than skipping over it.)

If you press F10 again, the debugger proceeds to the line that calls the function put_elogit_data_matrix in the ELogit_types module. Press F10 once more to find out whether this function was successful in placing the data matrix into the per-instance data. Look inside the derived type component OBJECTDATA%SESSION%DATASET%DATA_MATRIX, and it should appear as shown in Figure 8.6. Finally, to complete execution of the rest of the COM server code associated with the current line of VBA, press F5 in Visual Studio.

8.1.5   Finishing the Excel Client

Our Excel client does not yet display any results from the model fit. For this, we need to get some array-valued properties from the COM server. Retrieving an array from a COM server into Excel is a bit tricky because we must declare variant arrays and do some looping. We can't just tell Excel, "Put an array of numbers into spreadsheet range X" because we don't necessarily know in advance how large the range must be. When displaying array results to an Excel spreadsheet, it's better to do it in a cell-by-cell manner.

To implement the gets, first add these declarations near the top of the client code:

```
' Declarations
Dim i As Integer
Dim vBeta() As Variant
Dim vCovB() As Variant
Dim vBetaNames() As Variant
```

Before the code that releases the object, type the following lines. These lines display fit statistics and parameter estimates and do some simple calculations to display the standard errors and z-ratios.

```
' Get results
Range("Loglik").Value = objEL.loglik
Range("ChiSq").Value = objEL.X2
Range("GSq").Value = objEL.G2
Range("df").Value = objEL.df

' Load estimates into variant arrays
vBeta = objEL.beta
vCovB = objEL.cov_beta
vBetaNames = objEL.beta_names

' Loop through parameters, calculating
' standard errors and z-ratios
For i = 1 To UBound(vBeta, 1)
    Range("Beta").Cells(i) = vBeta(i, 1)
    Range("StdErr").Cells(i) = Sqr(vCovB(i, i))
    Range("Ratio").Cells(i) = vBeta(i, 1) _
    / Sqr(vCovB(i, i))
    Range("EstimLabel").Cells(i) = vBetaNames(i, 1)
Next i
```

We should round out this Excel client with some simple error handling. Just after the declarations section, insert the following lines:

```
' Set up error handling
```

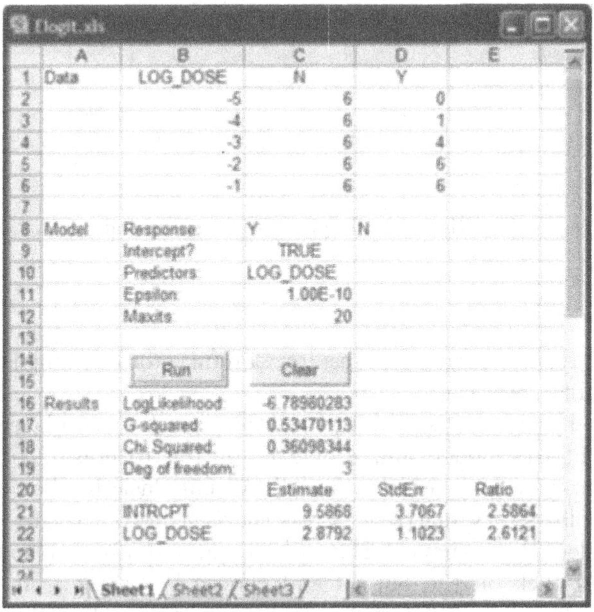

FIGURE 8.7. ELOGIT Excel client spreadsheet.

```
On Error GoTo ErrHandler
```

Then add these lines near the end of the script, after **Exit Sub** but before **End Sub**.

```
ErrHandler:
    ' Report error information from COM server
    If objEL.errMsgPresent Then
        Call MsgBox("Error: " & objEL.errMessage, vbCritical)
    Else
        Call MsgBox("Error: an unspecified error occurred." _
        & vbCrLf & Err.Description, vbCritical)
    End If
```

Try running the macro now. If all is well, you should see results appear in the appropriate cell ranges on Sheet1, as shown in Figure 8.7.

To test the error handling, change the predictor from LOG_DOSE to Y, and run the macro again. What happens? The error message that has been stored by calls to **err_handle** in the Fortran code should now emerge, indicating that Y cannot be used as a predictor (Figure 8.8). Notice that the error handler even reports the location in the Fortran source where the error was detected, which can be very useful to the developer. If this kind of error had not been trapped in the Fortran code, something undesirable would have happened. The COM server might have crashed or, worse yet,

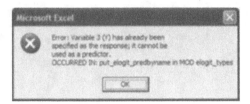

FIGURE 8.8. Simple error message dialog.

it may have reported erroneous results. Careful error handling is a hallmark of high-quality software.

As a final touch, let's add two buttons to the spreadsheet: a "run" button to fit the model, and a "clear" button to erase the results. To add buttons, you must activate Excel's Visual Basic toolbar by selecting View → Toolbars → Visual Basic. Press the ✴ icon to activate the controls toolbox. Select the Command Button tool and add a button to the spreadsheet. Right-click on the button and choose Properties from the popup menu. Change the name to cmdRun and the caption to Run. Then go to the Visual Basic editor window and add the following code in the editor. This is a new VBA subroutine and should be placed after the End Sub statement of Sub runELogit:

```
Private Sub cmdRun_Click()
    Call runELogit
End Sub
```

Add a similar button to clear the results. Name the button cmdClear, caption it Clear, and insert another subroutine.

```
Private Sub cmdClear_Click()
    Range("CaseID").ClearContents
    Range("Beta").ClearContents
    Range("StdErr").ClearContents
    Range("Ratio").ClearContents
    Range("EstimLabel").ClearContents
    Range("Loglik").ClearContents
    Range("ChiSq").ClearContents
    Range("GSq").ClearContents
    Range("df").ClearContents
End Sub
```

With these buttons, even a novice Excel user who knows nothing about macros or VBA can perform logistic regression.

8.1.6   Exercises

1. In our Excel spreadsheet, the physical arrangement of data and results works well for the viral assay dataset, but modifications will be needed to accommodate more cases or more predictor variables. Rearrange the spreadsheet so that it can handle larger datasets. Write a short document to show an Excel user how to apply the ELOGIT client to datasets of varying sizes.

2. Write a short document that explains to Excel users how to install the COM server and client on their own computers. Package the document together with the client and server so that the Excel user can install it. Test the procedure on a computer that has Excel but not Visual Studio.

3. Many diagnostics have been suggested for assessing the fit of a logistic regression model—for example, a scatterplot of residuals versus fitted values for the response (Hosmer and Lemeshow, 2000). Extend the ELOGIT COM server to provide on request the fitted values and Pearson residuals (see Equation 5.6). Then modify the client to display and plot the residuals versus the fitted values.

8.2   Clients for Other Environments

8.2.1   Keeping It Simple

In this section, we develop ELOGIT clients for a variety of computational packages used by statisticians and data analysts. Code for these clients differs in details and syntax, but the actions taken by each are essentially the same. Source-code files for each of our example clients are available on our Web site.

The main principle of client development is to keep it simple. All of the hard work should be done by the COM server. If you find yourself writing lots of client code, consider whether any of the actions being taken can be moved over to the server side.

8.2.2   Clients for S-PLUS and R

The S-PLUS client is one of our most straightforward. It does, however, use some undocumented features that allow us to pass arrays as properties. Before attempting to write a COM client in S-PLUS, you should first read the chapter on Automation in the online manual *S-PLUS 6 for Windows Programmer's Guide*. Specifically, you should familiarize yourself with these functions:

```
create.ole.object()
release.ole.object()
set.ole.property()
get.ole.property()
call.ole.method()
```

One nice feature of set.ole.property() and get.ole.property() is that they can put and get multiple properties at once, which helps us to keep the code short.

Many users of S-PLUS know little to nothing about the details of automation. For their benefit, we will handle those details by writing a function elogit() that simply accepts input data as arguments, fits the model and returns all the results as a list. A second function, elogit.print(), will take that list and generate a nicely printed summary.

Here is the code for elogit(). The argument x is a vector or matrix of predictors, and y is the vector of responses. The argument n should be a vector of the same length as y and containing the binomial denominators n_i; if this argument is missing, $n_i = 1$ is assumed.

_____ elogit.ssc _____
```
elogit <- function(x, y, n, intercept=T, maxits=20, eps=1D-08){

    # Create an elogit object instance
    pEL <- create.ole.object("ELogitSrvr.ELogitObj")
    if( is.null(pEL) ){
       stop("Unable to start ELOGIT COM server.")}

    # Coerce x into a matrix, if necessary
    if( is.vector(x) ){
       x <- matrix(x, ncol=1)}

    # Set up names for X-variables
    x.names <- dimnames(x)[[2]]
    if( is.null( x.names ) ){
       x.names_paste( "X", format(1:ncol(x)), sep="")}

    # Set up the data matrix, variable names and column numbers
    if( missing(n) ){
       data.matrix <- cbind(x,y)
       var.names <- c( x.names, "Y" )
       resp.col <- ncol(data.matrix)
       }
    else{
       data.matrix <- cbind(x,y,n)
       var.names <- c( x.names, "Y", "N" )
       resp.col <- c( ncol(data.matrix)-1, ncol(data.matrix) )
       }
    pred.col <- 1:ncol(x)

    # Load the data matrix, variable names and model specification
    result <- set.ole.property(pEL,
```

```
        list("data_matrix"=data.matrix, "var_names"=var.names,
        "response_bycol"=resp.col, "intercept"=intercept,
        "pred_bycol"=pred.col) )
  if( !all(result) ){
        err <- get.ole.property(pEL, "errMessage")$errMessage
        release.ole.object(pEL)
        stop(err)}

  # Fit the model
  result <- call.ole.method(pEL, "modelfit", maxits, eps)
  if( get.ole.property(pEL, "errMessagePresent")$errMessagePresent ){
        err <- get.ole.property(pEL, "errMessage")$errMessage
        release.ole.object(pEL)
        stop(err)
        }

  # Get results
  result <- get.ole.property(pEL,
        c( "iter", "converged", "beta", "beta_names", "cov_beta",
        "loglik", "X2", "G2", "df") )

  if( !result$converged ){
        warning( paste("Algorithm failed to converge by",
            format(result$iter), "iterations."))
        }

  # change underscore to period in list names
  names(result)[ names(result)=="cov_beta" ]_"cov.beta"
  names(result)[ names(result)=="beta_names" ]_"beta.names"

  # convert covbeta from a list of vectors into a square matrix
  result$cov.beta <- matrix(unlist(result$cov.beta),
        nrow=length(result$cov.beta))

  # Unload the eLogit object
  release.ole.object(pEL)

  # Return list of results
  result}
```

We need to point out two minor details. First, some of our COM object's properties have names containing underscore characters. A name containing an underscore causes problems for the S language because the underscore character is a deprecated form of the assignment (<-) operator. To avoid difficulties, names with embedded underscores have been enclosed by quotation marks. The second point is that when the COM server sends a rank-two array to S-PLUS, the get.ole.property function returns it as a list. To convert it into a matrix, you need to unlist and reshape it as shown above for the estimated covariance matrix.

Our second function, elogit.print(), accepts as its sole argument the list created by elogit.

```
┌──────────────────────────── elogit.ssc ────────────────────────────┐
│                                                                      │
│  elogit.print_function( result ){                                    │
│                                                                      │
│     #  print table of coefficients                                   │
│     coef <- result$beta                                              │
│     SE <- sqrt( diag( result$cov.beta ) )                            │
│     z <- coef/SE                                                     │
│     pval <- 2 * pnorm(-abs(z))                                       │
│     coef.table <- cbind( coef, SE, z, pval)                          │
│     dimnames( coef.table ) <- list(                                  │
│        result$beta.names, c("coef", "SE", "z", "pval") )             │
│     cat( "Summary of estimated coefficients:\n")                     │
│     print(coef.table)                                                │
│                                                                      │
│     #  print convergence summary                                     │
│     cat("\n")                                                        │
│     if( result$converged ){                                          │
│        cat( paste( "Algorithm converged in",                         │
│        format(result$iter), "iterations.", "\n"))                    │
│        }                                                             │
│     else{                                                            │
│        cat( paste( "Algorithm failed to converged by",               │
│        format(result$iter), "iterations.", "\n"))                    │
│        }                                                             │
│                                                                      │
│     #  print fit statistics                                          │
│     cat("\n")                                                        │
│     cat( "Summary of model fit:\n")                                  │
│     cat( paste(                                                      │
│        c("Loglikelihood =", "Pearson X2    =", "Deviance G2   ="),   │
│        format( c( result$loglik, result$X2, result$G2 ) )), sep="\n")│
│     cat( paste(                                                      │
│        "Degrees of freedom:", format(result$df), "\n"))              │
│                                                                      │
│     # return a non-printing NULL value                               │
│     invisible()                                                      │
│     }                                                                │
│                                                                      │
└──────────────────────────────────────────────────────────────────────┘
```

Here is an example S-PLUS command session that uses the client.

```
> # read the client code
> source("elogit.ssc")

> # enter viral assay data
> y <- c(0,1,4,6,6)         # number dead per group
> n <- c(6,6,6,6,6)         # group sizes
> x <- c(-5,-4,-3,-2,-1)    # log dose

> tmp <- elogit(x, y, n)
```

```
> elogit.print( tmp )
Summary of estimated coefficients:
          coef      SE        z        pval
INTRCPT 9.586808 3.706673 2.586364 0.009699431
    X1 2.879165 1.102251 2.612078 0.008999360

Algorithm converged in 8 iterations.

Summary of model fit:
Loglikelihood = -6.7898028
Pearson X2    =  0.3609834
Deviance G2   =  0.5347011
Degrees of freedom: 3
```

To create a COM client for R, you must install the `rcom` package as mentioned in Chapter 7. You will need to make minor adjustments to `elogit()` in the parts that call the COM server; refer to the `rcom` documentation for details.

8.2.3   A Client for SAS

Writing COM clients for SAS is a bit more challenging. The main difficulty is that, in the current version (Version 8), SAS provides COM functionality only through the SAS Component Language (SCL). Most users of SAS have never heard of SCL and may prefer not to learn it.

To simplify the task of writing COM clients in SAS, we have created a tool that allows you to invoke COM servers directly from an ordinary SAS program through PROC IML. This tool, which we call SASCOMIO (for "SAS/COM interoperability"), consists of a Windows DLL (`sascomio.dll`) and another file (`sascomio.cbt`) that defines the DLL calling conventions. Before proceeding with this current section, we suggest that you briefly review the material in Section 6.5.5 concerning DLLs in SAS.

A list of the procedures available in SASCOMIO is shown in Table 8.2. We use these procedures to invoke COM properties and methods in much the same way as in our previous client examples. In order to use the SASCOMIO library, you must install `sascomio.dll` and the associated `.cbt` file on your computer. The `.dll` file may be placed in any folder listed in the Windows PATH variable, or in the folder where other SAS DLLs are located (i.e., in the `core\sasexe` subfolder of the SAS installation). The `.cbt` file can be placed anywhere, as long as you provide the full file path in the `filename sascbtbl` directive near the beginning of the PROC IML call. We typically place the `.cbt` file in the same folder as our SAS client program.

TABLE 8.2. Functions in the SASCOMIO library.

Function	Description
CreateObj	Creates an instance of the object described by sProgID and returns a pointer to that object
ReleaseObj	Releases the object pointed to by lDispPtr (no return value)
PutPropNum PutPropChar	Puts a property and returns an hresult *Arguments:* lDispPtr (pointer to the object instance) sPropName (name of the property to put) dim1 and dim2 (dimensions of the array) Value (numeric/character value or array of values) nArgs (number of arguments) Arg1dim1 and Arg1dim2 (dimensions of the array) Arg1Value (character value or array of values) (Repeat if there are two arguments)
GetPropNum GetPropChar	Gets a property and returns an hresult *Arguments:* lDispPtr (pointer to the object instance) sPropName (name of the property to get) dim1 and dim2 (dimensions of the array) Value (numeric/character value or array of values) nArgs (number of arguments) Arg1dim1 and Arg1dim2 (dimensions of the array) Arg1Value (character value or array of values) (Repeat if there are two arguments)
RunMethod	Runs a method and returns an hresult *Arguments:* lDispPtr (pointer to the object instance) sMethName (name of the method to invoke) nArgs (number of arguments) sArg1Name (optional, name of first argument) Arg1dim1 and Arg1dim2 (dimensions of the array) Arg1Value (character value or array of values) (Repeat for each argument, up to four arguments)
errorLookup	Receives an hresult value and produces an error message; returns the number of characters in the message *Arguments:* lHresult (the hresult value to look up) sMessage (the corresponding error message)

Here is a SAS program that defines an IML module that invokes the COM server and passes data, model information, and results through the module arguments. The module is stored for later use in a SAS catalog called `mylib`.

```
─────────────── elogit.sas ───────────────
libname elogit '.';

proc iml;
  /* IML Module for Logistic Regression */
  start elogit(dataset, varnames, resp, inter, pred, eps, maxits,
    converged, iter, loglik, X2, G2, df, beta, betanames, covBeta,
    Err, errMessage);

    /* load SAS/COM interop function defs */
    filename sascbtbl 'sascomio.cbt';

    /* initializations */
    nkase = nrow(dataset);
    nvar = ncol(dataset);
    converged = 0;
    iter = 0.0;
    loglik = 0.0;
    G2 = 0.0;
    X2 = 0.0;
    df = 0.0;
    Err = 0;
    result = 0;

    /* Make enough room for 160 characters in errMessage */
    errMessage = '';
    do i=1 to 16;
      errMessage = errMessage + '          ';
    end;

    /* Determine the number of parameters */
    nparam = ncol(pred);
    if inter = 'True' then nparam = nparam + 1;

    /* Shape the beta, covBeta, and betanames arrays */
    beta = shape( 0.0, nparam, 1);
    betanames = shape( '        ', nparam, 1);
    covBeta = shape( 0.0, nparam, nparam);

    /* Create ELogit object instance */
    ProgID = 'ELogitSrvr.ELogitObj';
      pElogit = modulein('createobj', ProgID );

      if pElogit = 0 then do;
        /* Failed to create object -- return error */
        Err = -1;
        errMessage = "Error: Unable to load ELogit object.";
      end;
      else do;
```

```
    /* Load the data matrix and variable names */
result = modulein('putpropnum', pElogit, 'data_matrix',
  nkase, nvar, dataset, 0);
if result ^= 0 then goto err_hndl;
result = modulein('putpropchar', pElogit, 'var_names',
  1, nvar, varnames, 0);
if result ^= 0 then goto err_hndl;

/* Load the model specification */
result = modulein('putpropchar', pElogit, 'response_byname',
  1, 2, resp, 0);
if result ^= 0 then goto err_hndl;
result = modulein('putpropchar', pElogit, 'pred_byname',
  1, 1, pred, 0);
if result ^= 0 then goto err_hndl;
result = modulein('putpropchar', pElogit, 'intercept',
  1, 1, inter, 0);
if result ^= 0 then goto err_hndl;

/* Fit the model */
result = modulein('runmethod', pElogit, 'modelfit', 2,
  'maxits', 1, 1, char(maxits), 'eps', 1, 1, char(eps)  );
if result ^= 0 then goto err_hndl;

/* Get convergence info */
result = modulein('getpropnum', pElogit, 'converged',
  1, 1, converged, 0);
if result ^= 0 then goto err_hndl;
result = modulein('getpropnum', pElogit, 'iter',
  1, 1, iter, 0);
if result ^= 0 then goto err_hndl;

/* Get general statistics */
result = modulein('getpropnum', pElogit, 'loglik',
  1, 1, loglik, 0);
if result ^= 0 then goto err_hndl;
result = modulein('getpropnum', pElogit, 'G2',
  1, 1, G2, 0);
if result ^= 0 then goto err_hndl;
result = modulein('getpropnum', pElogit, 'X2',
  1, 1, X2, 0);
if result ^= 0 then goto err_hndl;
result = modulein('getpropnum', pElogit, 'df',
  1, 1, df, 0);
if result ^= 0 then goto err_hndl;

/* Get parameter estimates */
result = modulein('getpropnum', pElogit, 'beta',
  nparam, 1, beta, 0);
if result ^= 0 then goto err_hndl;
result = modulein('getpropnum', pElogit, 'cov_beta',
  nparam, nparam, covbeta, 0);
if result ^= 0 then goto err_hndl;
```

```
    result = modulein('getpropchar', pElogit, 'beta_names',
        nparam, 1, betanames, 0);
    if result ^= 0 then goto err_hndl;

    /* Get errors, if any */
err_hndl:
    if result ^= 0 then do;
        ret = modulein('getpropnum', pElogit, 'errMessagePresent',
            1, 1, Err, 0);
        if Err = -1 then do; /* -1 = true, 0 = false */
        /* Get internal error from COM Server */
            ret = modulein('getpropchar', pElogit, 'errMessage',
                1, 1, errMessage, 0);
        end;
        else do;
            /* Get error from windows system */
            Err = -1;
            ret = modulein('errorlookup', result, errMessage );
        end;
    end;

    /* Unload the ELogit object */
        call modulei('releaseobj', pElogit);
    end;
finish; /* elogit */

reset storage=elogit.mylib;
store module=elogit;
quit;
```

The advantage of setting up the client as an IML module is that the calls to the procedures in SASCOMIO, which are not part of the standard SAS language, are hidden inside the module; users won't need to know anything about SASCOMIO or even about COM.

A sample SAS program that applies the module to the viral assay data is shown below. Notice how this program uses the catalog file `mylib` which was created by running `elogit.sas`. Notice also that it is necessary to load the file `sascomio.cbt` again in this program before using the `elogit` module.

─────────── run_elogit.sas ───────────

```
libname elogit '.';

proc iml;
    /* Perform Logistic Regression on VIRAL dataset */
    /* Parameters */
    ver = 1.1;
    author = 'A. Programmer';

    /* load SAS/COM interop function defs */
    filename sascbtbl 'sascomio.cbt';
```

```
/* load elogit module */
reset storage=elogit.mylib;
load module=elogit;

/* load the dataset and variable names into arrays */
use elogit.viral;
read all into dataset [colname=varnames];

/* model info */
resp = { 'Y' 'N' };
pred = { 'LOG_DOSE' };
inter = 'True';

/* convergence info */
maxits = 20;
eps = 1.0E-10;

/* run logistic regression procedure */
run elogit(dataset, varnames, resp, inter, pred,
  eps, maxits, converged, iter, loglik, X2, G2,
  df, beta, betanames, covBeta, Err, errMessage );

if Err = 0 then do;

  /* Compute standard error */
  stdErr = sqrt(vecdiag(covBeta));

  /* Likelihood ratio */
  ratio = beta / stdErr;

  /* Display results */
  print 'elogit.sas','Logistic regression using the',
    'ELogit COM Server', '', 'Version ' (ver[1]),
    (author[1]);
  print 'Data set information',
    'Number of cases:' (nrow(dataset)),
    'Number of variables:' (ncol(dataset));
  print dataset[format=8.0];
  print 'Model specification',
    ({ 'Response (y):', 'Denominator (n):', 'Predictors:' })
    ( resp` // betanames );
  print 'Iteratively reweighted least-squares algorithm';
  if converged = -1 then do;
    print 'Converged at iteration: ' (iter[1]);
  end;
  else do;
    print 'Failed to converge by iteration: ' (iter[1]);
  end;
  headings = {'estimate' 'std.err' 'ratio'};
  print (betanames[,1]) (beta || stdErr || ratio)
    [format=8.4 colname={'estimate' 'std.err' 'ratio'}];
  print 'Summary of model fit',
    ({'Loglikelihood','G-squared','Chi Squared','Deg of freedom'})
```

```
     (loglik // G2 // X2 // df);
  end;
  else do;
    /* Print error message */
    print errMessage;
  end;

quit;
```

The output from our SAS program looks like this:

```
                        elogit.sas
                 Logistic regression using the
                     ELogit COM Server

                    Version        1.1
                    A. Programmer


                    Data set information
                  Number of cases:        5
                  Number of variables:         3


                         DATASET
            LOG_DOSE         N          Y
                 -5          6          0
                 -4          6          1
                 -3          6          4
                 -2          6          6
                 -1          6          6


                    Model specification
                  Response (y):    Y
                  Denominator (n): N
                  Predictors:      INTRCPT
                                   LOG_DOSE


          Iteratively reweighted least-squares algorithm


                  Converged at iteration:         8

                       estimate std.err  ratio
             INTRCPT    9.5868   3.7067  2.5864
             LOG_DOSE   2.8792   1.1023  2.6121


                     Summary of model fit
                   Loglikelihood  -6.789803
```

```
G-squared       0.5347011
Chi Squared     0.3609834
Deg of freedom         3
```

8.2.4 SPSS

An SPSS client can be implemented as an SPSS macro or script. These
macros are written in a language that closely resembles VBA. Our script is
set up as a function called from SPSS syntax code, which is more familiar
to SPSS users than the script language. The code for this client is longer
than the code for our VBA client in Excel. The extra length is needed to
extract data from the SPSS dataset (shown in Figure 8.10 and to set up
the model information passed as arguments to the script.

———————————— ELogit.sbs ————————————

```
Option Explicit
'#Uses "ELogitOutput.sbs"
Sub Main
    ' Script that uses the ELogit COM Server to fit
    '     a logistic regression model using the currently
    '     active data set and the parameters specified.
    '
    ' Parameters:
    '     1) Number of response variables (if greater
    '          than 1, it indicates that there is a
    '          denominator to the response variable,
    '          and 2 "response"
    '          variables are given, the second being
    '          the denominator variable of the intercept.
    '     2) Number of predictors
    '     3) Response variables
    '     4) Predictor variables
    '     5) Intercept (True or False)
    '     6) Convergence criterion
    '     7) Maximum iterations
    '
    '     Example: "2 1 y n log_dose True 1.0e-9 20"

    ' Declare local variables
    Dim sDataFile As String
    Dim vTemp As Variant
    Dim vData() As Variant

    Dim nVars As Integer
    Dim nKase As Integer

    Dim i As Integer
    Dim j As Integer

    Dim sParam As String
```

```
Dim sParams() As String
Dim vResp() As Variant
Dim vPred() As Variant
Dim ListVars() As String
Dim nResp As Integer
Dim nPred As Integer
Dim bInter As Boolean
Dim eps As Double
Dim maxits As Integer

' Declare data document object from SPSS object library
Dim objDataDoc As ISpssDataDoc

' Declare ELogit COM Server object
Dim objEL As Object

' Set up error handling
On Error GoTo ErrHandler

' Instantiate the ELogit COM Server object
Set objEL = CreateObject("ELogitSrvr.ELogitObj")

'  Obtain object pointing to the currently open data document
Set objDataDoc = objSpssApp.Documents.GetDataDoc(0)

' Get the parameter string
' First, see if it is being passed by a syntax SCRIPT directive
'sParam = "2 1 y n log_dose True 1.0e-9 20" '(for debugging)
sParam = objSpssApp.ScriptParameter(0)
If sParam = "" Then
    ' Second, see if it is being called from another macro
    sParam = Command$
    If sParam = "" Then
        Err.Raise -1,,"No arguments to script call." & vbCrLf & _
            "Please check syntax file or script code.",,
        Exit Sub
    End If
End If

' Extract parameters from string
i = 1
Do
    ' Build an array of the space-delimited parameters
    ReDim Preserve sParams(1 To i)
    ' Look for first occurrence of space delimiter
    j = InStr(1,sParam," ")
    If j > 0 Then
        ' Extract left side as a parameter
        sParams(i) = Left( sParam, j-1 )
        ' Continue searching the right side
        sParam = Right( sParam, Len(sParam)-j )
        i = i + 1
    Else
```

```
            ' Last parameter was found
            sParams(i) = sParam
            Exit Do
        End If
Loop

' Load parameters into appropriate variables
nResp = CInt( sParams(1) )
nPred = CInt( sParams(2) )
ReDim vResp(1 To nResp)
ReDim vPred(1 To nPred)
i = 3
For j = 1 To nResp
    vResp(j) = sParams(i)
    i = i + 1
Next j
For j = 1 To nPred
    vPred(j) = sParams(i)
    i = i + 1
Next j
If sParams(i) = "True" Then
    bInter = True
Else
    bInter = False
End If
i = i + 1
eps = CDbl( sParams(i) )
i = i + 1
maxits = CInt( sParams(i) )

' Get the variable names, nVars, nKase,
' and data file name from the SPSS data file
ListVars = objDataDoc.GetVariables(False)
nVars = objDataDoc.GetNumberOfVariables
nKase = objDataDoc.GetNumberOfCases
sDataFile = objDataDoc.GetDocumentPath

' Get data matrix
vTemp = objDataDoc.GetTextData(ListVars(0),ListVars(nVars-1),1,nKase)

' Release the data document object -- we're done with it.
Set objDataDoc = Nothing

' Dynamically allocate data matrix
' In SPSS, arrays are indexed from zero
ReDim vData(0 To nKase-1, 0 To nVars-1)

' Transpose data matrix (COM server needs row-major)
For i = 0 To nVars-1
    For j = 0 To nKase-1
        vData(j,i) = vTemp(i,j)
    Next j
Next i
```

```
' Load data matrix and variable names into COM server
objEL.data_matrix = vData
objEL.var_names = ListVars

' Load the model info
objEL.response_byname = vResp
objEL.pred_byname = vPred
objEL.intercept = bInter

' Fit model
Call objEL.modelfit(maxits, eps)

' Generate an SPSS output document
Call ELogitOutput( nKase, nVars, _
    sDataFile, vData, vResp, vPred, _
    objEL.converged, objEL.iter, _
    objEL.beta, objEL.cov_beta, objEL.beta_names, objEL.loglik, _
    objEL.G2, objEL.X2, objEL.df, ListVars, bInter )

' Release the ELogit COM Server object
Wait 1
Set objEL = Nothing
' Wait a second for object to close (tends to crash if you don't)
Wait 1

Exit Sub
ErrHandler:
    ' Error handler.  Display the error in a dialog box.
    Call MsgBox(Err.Description & ": " & objEL.errMessage)
    Set objEL = Nothing
    Wait 1

End Sub
```

Another SPSS script file, which we do not show here, defines a function ELogitOutput for printing results to an output file. Both of these files can be downloaded from our Web site.

Here is an example SPSS program written in syntax code that runs the ELogit script. Even SPSS users who are unaccustomed to syntax code will quickly get the idea:

────────────────── ELogitSyntax.sps ──────────────────
```
SCRIPT file="elogit.sbs"
("2 1 y n log_dose True 1.0e-10 20") .
```

For users who prefer to analyze their data through menu-driven dialogs, we have written another SPSS script that provides a graphical dialog for ELOGIT. The dialog window, shown in Figure 8.9, allows the user to interactively specify the response and predictor variables before fitting the

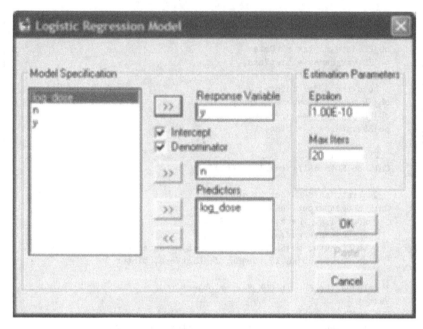

FIGURE 8.9. SPSS dialog interface for ELOGIT.

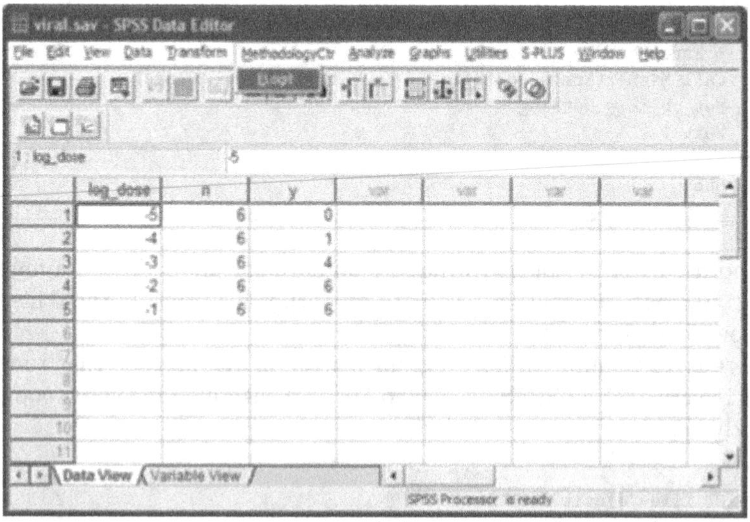

FIGURE 8.10. SPSS menu item for launching the ELOGIT dialog script.

model. The SPSS menu can be customized to include an item for this script, as shown in Figure 8.10. Source code for the dialog script is available from our Web site.

8.2.5 MATLAB

Invoking COM servers is very straightforward in MATLAB. Our ELOGIT
client for MATLAB, which has been set up as a MATLAB script or M-file,
is shown below.

```
——————————————————— elogit.m ———————————
function [loglik, x2, g2, df, Beta, betaNames, CovBeta, ...
        converged, iter, err ] ...
    = elogit(data, varNames, resp, inter, pred, eps, maxits)

% Begin try/catch
err = '';
try

        % Create an elogit object instance
        h = actxserver( 'ELogitSrvr.ELogitObj');

        % Load the data matrix and column labels
        set(h, 'data_matrix', data);
        set(h, 'var_names', varNames);

        % Load the model specification
        set(h, 'response_byname', resp);
        set(h, 'intercept', inter);
        set(h, 'pred_byname', pred);

        % Fit the model
        invoke( h, 'modelfit', maxits, eps );

        % Get convergence info
        converged = get(h, 'converged');
        iter = get(h, 'iter');

        % Get general statistics
        loglik = get(h, 'loglik');
        g2 = get(h,'g2');
        x2 = get(h,'x2');
        df = double(get(h,'df'));

        % Get parameter estimates
        Beta = get(h, 'beta');
        CovBeta = get(h,'cov_Beta');
        betaNames = get(h,'beta_Names');


catch %Error handler

    if get(h, 'errMsgPresent')
        err = get(h,'errMessage');
    else
        err = lasterr;
    end
```

```
end

% Release the COM Server object
release(h);
```

Notice how errors are handled through the `try...catch` construct. If you are unfamiliar with `try...catch`, a brief introduction will be given in the next section. Here is another script that applies ELOGIT to the viral assay data and prints results:

—————————— run_elogit.m ——————————

```
% Clear workspace
clear all;

% Parameters
ver = 1.1;
author = 'David R. Lemmon';

% Create the data array (viral dataset)
nvar = 3;
nkase = 5;
data = zeros(nkase, nvar);
data(:,1) = (-5:-1)';
data(:,2) = 6;
data(:,3) = [ 0; 1; 4; 6; 6 ];

% Create data labels cell array
varNames = cellstr(['LOG_DOSE'; 'N       '; 'Y       '])';

% Create model
%    Response
resp = cellstr(['Y';'N'])';
%    Intercept
inter = logical(1);
%    Predictors
pred = 'LOG_DOSE';

% Fit model
eps = 1.e-10;
maxits = 20;
[loglik, x2, g2, df, Beta, betaNames, CovBeta, converged, iter, err ] ...
    = elogit(data, varNames, resp, inter, pred, eps, maxits);

% Check for errors
if err ~= ''
    fprintf('Error: %s', err);
else

        % Get size of Beta
        [szBeta,junk] = size(Beta);

        % Compute standard error
        stdErr = sqrt(cat(1,CovBeta{logical(eye(szBeta))}));
```

```
        % Compute z-ratio
        aBeta = cat(1,Beta{:});
        ratio = aBeta ./ stdErr;

        % Display results
        fprintf('\nELogit.m\nLogistic Regression using');
    fprintf('the ELogit COM Server\n');
        fprintf('Version %4.1f\n%s\n',ver,author);
        %     Date and time
        fprintf('\n%s\n',datestr(now));
        %     Dataset info
        fprintf('\nData set information\n');
        fprintf('\tNumber of cases:      %5d\n',double(nkase));
        fprintf('\tNumber of variables: %5d\n',double(nvar));
        fprintf('\n\tVariables\n\t---------------\n');
        for i=1:double(nvar)
        fprintf('%8d\t%s\n',i,varNames{i});
        end
        %     Model
        fprintf('\nModel specification\n');
        fprintf('\tResponse (y):   \t%s\n',resp{1});
        fprintf('\tDenominator (n):\t%s\n',resp{2});
        for i=1:szBeta
        if i==1
            fprintf('\t%s\t%s\n','Predictors:      ',betaNames{i});
        else
            fprintf('\t%s\t%s\n','                  ',betaNames{i});
        end
        end
        %     Convergence info
        fprintf('\nIteratively reweighted least-squares algorithm\n');
        if converged
        fprintf('\tConverged at iteration %d\n',double(iter));
        else
        fprintf('\tFailed to converge by iteration %d\n',double(iter));
        end
        %     Estimates
        fprintf('\n                     estimate     std.err.     ratio');
        fprintf('\n                     ----------- ---------- -----------\n');
        for i=1:szBeta
        fprintf('\t%s  \t%7.4f\t\t%7.4f\t\t%7.4f\n',betaNames{i},...
            Beta{i},stdErr(i),ratio(i));
        end
        %     Model fit
        fprintf('\nSummary of model fit\n');
        fprintf('\n\tLoglikelihood:\t%11.8f\n',loglik);
        fprintf('\tDeviance G^2:\t%11.8f\n',g2);
        fprintf('\tPearson''s X^2:\t%11.8f\n',x2);
        fprintf('\tDeg. of freedom:\t%d\n',df);
end
```

8.3 Creating a Standalone GUI Application

8.3.1 Component-Based Design

If you are not a professional programmer, the idea of creating a statistical application with its own graphical user interface may seem daunting. With modern development tools and languages such as Visual Basic, it is actually quite easy to arrange menus, buttons, checkboxes, and so on into a GUI with a professional look and feel. Integrating graphical components with the computational procedures may be tricky, however, especially if those procedures and the GUI are written in different languages.

Fortunately, if you have implemented your computational procedures as a COM server, connecting that server to a user-friendly GUI can be a realizable goal, even if time and resources are limited. We will demonstrate the entire process by creating a graphical, stand-alone version of ELOGIT, which we call ELogit.NET.

When embarking on a project such as this, we strongly recommend that you implement the computational procedures in a COM server first. It is extremely difficult to create a GUI when the properties and methods of your COM server do not yet exist. But if your server is up and running, the GUI—which is just another type of COM client—can be tested continually and thoroughly throughout the development process. In fact, the properties and methods of the COM server will actually drive the GUI's design. With a working COM server in hand, you can also build up your GUI incrementally. It can start as a bare-bones tool for getting data into the COM server and shipping the results out. Over time, fancier elements can be added as the whole interface takes shape.

8.3.2 Visual Basic .NET

In this example, we will create our GUI using Visual Basic .NET. This language, which replaced Visual Basic in 2001, is relatively easy to learn and is a standard industry tool for creating Windows-based applications. A large number of reference books are available to help you learn about Visual Basic .NET and about the .NET platform on which it is based. The .NET platform is compatible with COM. In Visual Basic .NET, you can create instances of COM objects and invoke their methods and properties just as we did in our client for Excel. Using the tools provided by Microsoft Visual Studio .NET, adding windows, buttons, and other visual elements is simply a matter of dragging and dropping them into the project.

8.3.3 An XML Data Format

With ELogit.NET, a user will be able to interactively open a data file, specify a model, run the model fit procedure, and view and save the results.

Therefore, we must first decide on a format for the data file. When we wrote the ELOGIT console application back in Chapter 5, we developed a system for reading a rectangular data matrix and variable names from ordinary text or ASCII files. Methods for reading the data and names files are already available in our COM server, and using those methods would certainly be a reasonable approach. To illustrate a new trend in programming, however, we will show you how to read a data matrix that has been stored in a modern format called XML.

XML, an abbreviation for Extensible Markup Language, is a descendant of Hypertext Markup Language (HTML). Whereas HTML was used primarily for creating Web pages, XML is a more general format for storing tabular data and all kinds of information in text files in a self-documenting fashion. The contents of an XML file can be read directly into applications such as Excel and programs written in new languages such as Visual Basic .NET. To see what XML looks like, here is our viral assay dataset in XML.

```
──────────────── viral.xml ────────────────
<?xml version="1.0" encoding="windows-1252" ?>
<TABLE>
   <VIRAL>
       <LOG_DOSE> -5 </LOG_DOSE>
       <N> 6 </N>
       <Y> 0 </Y>
   </VIRAL>
   <VIRAL>
       <LOG_DOSE> -4 </LOG_DOSE>
       <N> 6 </N>
       <Y> 1 </Y>
   </VIRAL>
   <VIRAL>
       <LOG_DOSE> -3 </LOG_DOSE>
       <N> 6 </N>
       <Y> 4 </Y>
   </VIRAL>
   <VIRAL>
       <LOG_DOSE> -2 </LOG_DOSE>
       <N> 6 </N>
       <Y> 6 </Y>
   </VIRAL>
   <VIRAL>
       <LOG_DOSE> -1 </LOG_DOSE>
       <N> 6 </N>
       <Y> 6 </Y>
   </VIRAL>
</TABLE>
```

Readers familiar with HTML will immediately notice many similarities. XML has beginning and ending tags enclosed in angle braces, with a forward slash to designate an ending. There are also attributes inside tags, such as the **version=** designation. Unlike HTML tags, however, the tags in

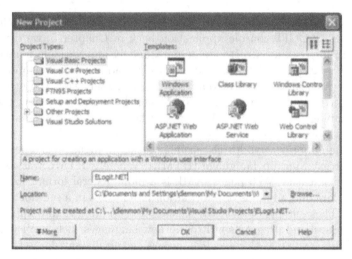

FIGURE 8.11. The New Project dialog.

XML do not specify how the file contents should be formatted in a browser window. Therefore, if you open up this XML file in a browser, it may not look any different from the way it is shown here.

How will users of ELogit.NET put their data into the XML format? We generated `viral.xml` from a SAS dataset in this manner:

```
libname viral '.';
libname out xml 'viral.xml';

proc copy in=viral out=out;
    select viral;
run;
```

XML files can also be created by Excel and many other programs. The fact that SAS can import and export XML files illustrates how XML is rapidly becoming a new industry standard for storing, sharing, and transmitting data.

8.3.4   Starting the Graphical Interface

To get started building the GUI, open up Microsoft Visual Studio .NET. When the Start window appears, press the "New Project" button. From among the types of Visual Basic projects shown in the New Project dialog box, select Windows Application. Enter the project name, ELogit.NET, as shown in Figure 8.11. We are now looking at a blank Windows form called Form1.vb. This form has a graphical front end with source code underlying it. Our job is to add graphical elements to the form and then fill in the

FIGURE 8.12. DataGrid and Button added to the form.

corresponding source code that tells the program what to do when various events (mouse clicks, keystrokes, etc.) occur.

8.3.5   Reading the Data File

The first feature of ELogit.NET that we will implement is to open and read the data file. ELogit.NET will read a dataset from an XML file and display it in a spreadsheet-like grid. In the Windows Forms Toolbox on the left, select DataGrid and draw the data grid on the form. Next, in the Properties box on the right, find the Dock property and change its value to Top by pressing the dropdown button and then pressing the topmost button. Change the value of the Size property to 292,136. Then select Button from the Toolbox and draw a button beneath the DataGrid. Change the Name property to btOpen, the Text property to Open Dataset, and the Size property to 96,24. Your form should then look like Figure 8.12.

Next, select the OpenFileDialog control from the Toolbox and add it by clicking anywhere on the form. An icon will then appear in a space below the form window. The purpose of this step will soon be apparent.

Now we must add event-handling code to make ELogit.NET do two things. First, when a user presses the "Open Dataset" button, a standard Windows file-selection dialog box will appear to allow the user to select the XML file to be opened. Second, after the XML file has been selected, the data from the file will be read into the program and displayed in the DataGrid. To add this code, move your mouse cursor to the form and double-click on the "Open Dataset" button. The form's code window will appear, and you will immediately see the skeleton of a subroutine named btOpen_Click. This is the code that will run when the button is pressed. Add the following code inside that subroutine.

```
Dim filePath As String
Dim column As DataColumn
Dim dsData As DataSet
 ' Show dialog to get path of xml data file
With OpenFileDialog1
    .Filter = "XML files (*.xml)|*.xml|" & _
        "Text files (*.txt)|*.txt|All files (*.*)|*.*"
    .DefaultExt = "*.xml"
    .ShowDialog()
    filePath = .FileName
End With

If filePath <> "" Then
    ' Read XML dataset
    dsData = New DataSet()
    With dsData
        .Reset()
        .ReadXml(filePath)
    End With
    ' View it in the data grid
    With DataGrid1
        .DataSource = dsData
        .DataMember = dsData.Tables(0).TableName
        .CaptionText = filePath
    End With
End If
```

What does this code do? Consider the part just below the declarations, which pertains to the Windows file-opening dialog. OpenFileDialog1 is a single instance of the OpenFileDialog object class that you added to this form. The With construct allows you to invoke its properties and methods without having to repeatedly type its name. The Filter and DefaultExt properties select the type of files to display when the window first appears. The ShowDialog method displays the window. After the selection has been made, the window is closed and the full path to the selected file is stored in the string variable filePath.

The second half of this code pertains to reading, storing, and displaying the data. In the .NET programming environment, tabular data may be stored in a DataSet object. We've declared one instance of a DataSet object, dsData, as a local variable in the button-click event procedure. The DataSet object class has an extremely handy method called ReadXml that, as the name implies, reads an entire XML file into the object. Once the XML data have been read into dsData, we can tell our DataGrid to display the data like this:

```
DataGrid1.DataSource = dsData
```

FIGURE 8.13. Data displayed in the DataGrid.

(Our code does exactly the same thing but within a **With** construct.) Because a DataSet object may contain more than one table, we must set the DataGrid's **DataMember** property to tell the grid which table to display. In this case, we point it to **dsData.Tables(0).TableName**, which directs the grid to display table zero; this is the only table stored in the XML file.

Now let's test this code. Press the Visual Studio .NET run button ▶ or the **F5** keyboard key. The project will automatically be compiled and then executed. When the form window appears, press the "Load Dataset" button; notice the effect of having specified **.xml** as the default file extension. Browse for **viral.xml** and open it. The viral assay data will then appear in the data grid, as shown in Figure 8.13. If you have another dataset available in XML format, try pressing the "Load Dataset" button again and tell ELogit.NET to load the new dataset. If everything works properly, the new dataset should replace the old one and appear in the DataGrid.

8.3.6 Specifying the Model

We must now provide a way for the user of ELogit.NET to select the data file. Switch back to the "Form view" by selecting the window tab **Form1.vb [Design]**. The first thing to do is to add more room. Change the **Size** property of Form1 to **360,590**. Then add each of the controls listed in Table 8.3. Because the GroupBox control is a container for all of the other controls, be careful to add the other controls within the confines of the GroupBox. Set the GroupBox property **Enabled** to **False** so that it will be disabled when the program first starts, along with the controls inside it. If you have arranged everything properly, the controls should appear as shown in Figure 8.14.

Whenever a new dataset is read from the XML file and displayed in the DataGrid, we want the names of the variables to be automatically

TABLE 8.3. Controls for model specification.

Object	Name	Label	Location	Size
GroupBox	gbModel	"Model Specification"	8,176	240,176
ListBox	lbVariables	—	8, 32	80,134
ListBox	lbResp	—	136, 32	96, 17
ListBox	lbDenom	—	136, 88	96, 17
ListBox	lbPred	—	136,128	96, 43
Label	lblVariables	"Variables"	8, 16	80, 16
Label	lblResp	"Response Variable"	128, 16	104, 16
Label	lblPred	"Predictors"	136,112	96, 16
Button	btResp	">>"	96, 32	32, 16
Button	btDenom	">>"	96, 88	32, 16
Button	btAddPred	">>"	96,128	32, 16
Button	btRemPred	"<<"	96,152	32, 16
CheckBox	cbInter	"Intercept"	112, 56	112, 16
CheckBox	cbDenom	"Denominator"	112, 72	112, 16

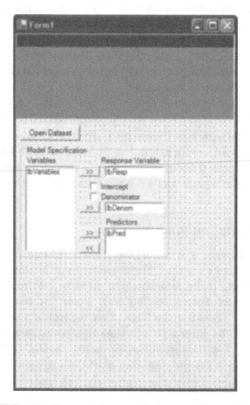

FIGURE 8.14. Form with controls for model specification.

displayed in the Variables ListBox. At the same time, we want the other model specification controls to be reset to sensible default values. To do this, add the following code to the btOpen_Click subroutine, just before the last End If statement:

```
' Load variable names into list box
With lbVariables
    .Items.Clear()
    For Each column In dsData.Tables(0).Columns
        .Items.Add(column.ColumnName)
    Next
End With

' Enable model group box
gbModel.Enabled = True

' Clear the other items in the model group
lbResp.Items.Clear()
lbDenom.Items.Clear()
lbPred.Items.Clear()
cbInter.Checked = False
cbDenom.Checked = False
btDenom.Enabled = False
lbDenom.Enabled = False
```

In this code, notice that column is a local instance of the object class DataColumn. It is capable of holding the data contained in one column of a table in a DataSet object.

Now let's add event handlers for some of the new controls by double-clicking them and adding code. The code for the button is btResp,

```
' Put selected item in the list box
lbResp.Items.Clear()
lbResp.Items.Add(lbVariables.SelectedItem)
```

for the button btDenom

```
' Put selected item in the list box
lbDenom.Items.Clear()
lbDenom.Items.Add(lbVariables.SelectedItem)
```

for the button btAddPred

```
' Add selected item to the list box
lbPred.Items.Add(lbVariables.SelectedItem)
```

for the button btRemPred

TABLE 8.4. Controls for model fitting-procedure.

Type	Name	Label or Value	Location	Size
GroupBox	gbEstim	"Estimation"	256,176	88, 96
TextBox	tbEps	1.0E-10	16, 32	64, 20
NumericUpDown	updMaxits	20	16, 72	64, 20
Label	lblEps	"Epsilon"	16, 16	56, 16
Label	lblMaxits	"Max Iters"	16, 56	56, 16
Button	btRun	"Run Estimation"	256,280	88, 32
DataGrid	DataGrid2	—	0,364	352,192

```
' Remove selected item from the list box
lbPred.Items.Remove(lbPred.SelectedItem)
```

and for the CheckBox cbDenom:

```
' Enable or disable denominator
If cbDenom.Checked Then
    btDenom.Enabled = True
    lbDenom.Enabled = True
Else ' unchecked
    btDenom.Enabled = False
    lbDenom.Enabled = False
End If
```

Compile and run the program again. Open a dataset and see how the form responds. Try adding and removing variables by selecting them from the listbox and pressing the buttons labeled ">>" and "<<."

8.3.7 Invoking the Model Fit Procedure

We are now ready to add a button that fits the specified model and add controls for the maximum number of iterations and convergence criterion. The results will be displayed in another DataGrid. Return to the form and add the controls listed in Table 8.4. Set the DataGrid's Dock property to Bottom. The form should now look like Figure 8.15.

Before continuing, we need to add a reference to our ELOGIT COM server. From the Visual Studio .NET menu, choose Project → Add Reference. In the dialog box, choose the COM tab and highlight ELogitObj 1.0 Type Library. Press Select and then OK, as shown in Figure 8.16. When you do this, Visual Studio .NET creates an interface library for the ELOGIT COM server so that it can be easily accessed by Visual Basic .NET. If you look in the obj subdirectory of the project folder, you will find this library in

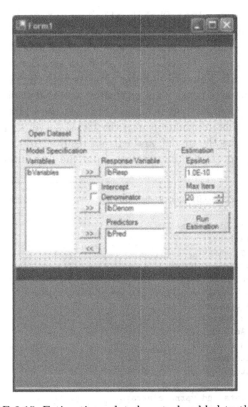

FIGURE 8.15. Estimation-related controls added to the form.

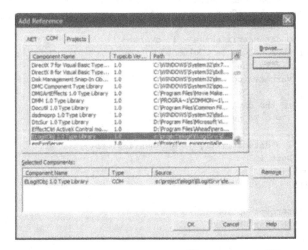

FIGURE 8.16. Adding a reference to the ELOGIT COM server.

a file called `Interop.ElogitObjLib.dll`. The library serves as the bridge between COM and .NET.

Now let's write the event-handling code that will execute when the "Run Estimation" button is pressed. Go to the form and double-click the "Run Estimation" button. The code window appears with the skeleton of the procedure called `btRun_Click`. This procedure will extract information from the user interface and arrange it in the form that the COM server is expecting. It will lead this information into the COM server's properties, invoke the method that fits the model, and extract results. Add the following code to `btRun_Click` to extract information from the user interface:

```
Dim objEL As ELogitObjLib.ELogitObj
Dim dsData As DataSet
Dim column As DataColumn
Dim nkase As Integer
Dim nvars As Integer
Dim datamatrix As Double(,)
Dim varnames As String()
Dim pred As String()
Dim resp As String()
Dim inter As Boolean
Dim stdErr As Double()
Dim ratio As Double()
Dim i As Integer
Dim j As Integer

' Form the data and varname arrays
dsData = DataGrid1.DataSource
With dsData.Tables(0)
    nvars = .Columns.Count
    nkase = .Rows.Count
    ReDim varnames(nvars - 1)
    ReDim datamatrix(nkase - 1, nvars - 1)
    For Each column In .Columns
        ' form the varname array
        varnames(column.Ordinal) = column.ColumnName
    Next
    ' form the data array
    For j = 0 To nvars - 1
        For i = 0 To nkase - 1
            datamatrix(i, j) = .Rows(i).Item(j)
        Next i
    Next j
End With

' Get intercept option
inter = cbInter.Checked

' Form response variable array
If inter Then
    ReDim resp(1)
```

```
        resp(0) = lbResp.Items(0)
        resp(1) = lbDenom.Items(0)
Else
    ReDim resp(0)
    resp(0) = lbResp.Items(0)
End If

' Form predictors array
ReDim pred(lbPred.Items.Count - 1)
For i = 0 To UBound(pred)
    pred(i) = lbPred.Items(i)
Next
```

Let's pause briefly to discuss this code. First, we create a local pointer called dsData and point it to the dataset in DataGrid1. Then, working with table zero (which, we presume, is the only table) in that dataset, we get the ColumnName property of the Columns object and store it in the local string array varnames. Next, we extract the data values from each row of the table (the Items property of each Row object in the table's Rows collection) and store it in the corresponding row of the local array datamatrix. Then we get the model information—the intercept option, response variable, denominator (if any), and predictors—by invoking the Items property of the ListBox objects and the Checked property of the CheckBox.

The next section of code uses a try...catch block, which is Visual Basic .NET's new way of handling errors. It is similar to the old on error goto statement but is more robust and aesthetically pleasing. This is how it works. First, try attempts to execute all of the statements placed below it. If any of those statements produces an exception, the point of execution immediately jumps to the section below catch. In this construct, there is also a section called finally that contains statements to be executed regardless of whether an exception occurred. Here is the code that loads the dataset and model into the ELOGIT COM server and invokes the model-fitting procedure. This block of code should be placed immediately below the previous block.

```
Try
    ' Create object instance
    objEL = New ELogitObjLib.ELogitObj
    ' Load the data matrix and column labels
    objEL.data_matrix = datamatrix
    objEL.var_names = varnames
    ' Load the model specification
    objEL.response_byname = resp
    objEL.intercept = inter
    objEL.pred_byname = pred

    ' Fit the model
```

```
    Call objEL.modelfit(updMaxits.Value, CDbl(tbEps.Text))

    ' Check convergence
    If objEL.converged Then
        MsgBox("Algorithm converged in " & objEL.iter & _
        " iterations.")
    Else
        MsgBox("Algorithm did not converge by " & _
        objEL.iter & " iterations.")
    End If

Catch ' Error handling
    If objEL.errMessagePresent Then
        'Com server internal error
        MsgBox("Error in COM server: " & vbCrLf & _
            objEL.errMessage, MsgBoxStyle.Critical)
    Else
        'Other errors
        MsgBox("Error " & Err.Number & ": " & _
            Err.Description, MsgBoxStyle.Critical)
    End If

Finally
    ' Clear the ELogit object
    objEL = Nothing

End Try
```

Compile and run the program again. Enter a proper model specification as shown in Figure 8.17, and then press the "Run Estimation" button. A dialog should appear to confirm that the algorithm converged in eight iterations. Now try to misspecify the model by adding the variable Y to the predictor variables. What happens? When you press "Run Estimation," the COM server provides a useful message: "Variable 3 (Y) has already been specified as the response; it cannot be used as a predictor....." This demonstrates that our try...catch block is working.

8.3.8 Displaying the Results

Now let's write some code to display the results. This code will also reside in the btRun_Click procedure, just before the Catch section. These statements will compute standard errors and z-ratios.

```
    ' compute standard error and z-ratio
    ReDim stdErr(UBound(objEL.beta))
    ReDim ratio(UBound(objEL.beta))
    For i = 1 To UBound(objEL.beta)
        stdErr(i) = Math.Sqrt((objEL.cov_beta(i, i)))
```

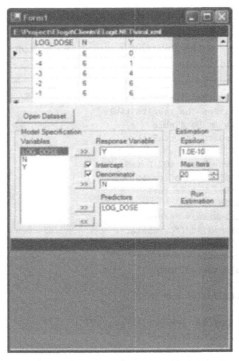

FIGURE 8.17. Running ELogit.NET with a proper model specification.

```
        ratio(i) = objEL.beta(i, 1) / stdErr(i)
    Next
```

Notice how we use the **beta** property from the ELOGIT COM server. Although **beta** is one-dimensional, it must be treated as a two-dimensional array in which the second dimension is one. The .NET framework always retrieves a COM server's variant array properties as two-dimensional arrays.

Now we will create a dataset object containing tables for holding the results and then display these results in **DataGrid2**. To do this, we need to declare a few more variables at the top of the procedure.

```
Dim dsResults As DataSet
Dim table As DataTable
Dim row As DataRow
```

The DataSet object **dsResults** will contain three tables—**estimates**, **fit_stats**, and **covBeta**—to hold the parameter estimates, fit statistics, and covariance matrix, respectively. We need to write some code to set up and initialize these tables within **dsResults**. Our **btOpen_Click** procedure is getting a bit long, so let's create another procedure and call it from **btOpen_Click**. Place the following code at the bottom of the form code file, just before **End Class**.

```
Private Sub initResultsTables()
    ' Initializes the results table
    Dim table As DataTable
    Dim dsResults As DataSet

    dsResults = DataGrid2.DataSource
    ' table of beta estimates
    table = New DataTable()
    With table
        .TableName = "estimates"
        .Columns.Add("parameter")
        .Columns.Add("estimate")
        .Columns.Add("std.err")
        .Columns.Add("ratio")
    End With
    dsResults.Tables.Add(table)

    ' table for fit statistics
    table = New DataTable()
    With table
        .TableName = "fit_stats"
        .Columns.Add("item")
        .Columns.Add("value")
    End With
    dsResults.Tables.Add(table)

    ' table for covariance matrix
    table = New DataTable()
    With table
        .TableName = "covBeta"
        .Columns.Add("row")
        .Columns.Add("column")
        .Columns.Add("value")
    End With
    dsResults.Tables.Add(table)
End Sub
```

Now go back to **btOpen_Click** and place this code just before the **Catch** section.

```
    ' Initialize the results table
    dsResults = New DataSet()
    DataGrid2.DataSource = dsResults
    Call initResultsTables()

    ' Extract results and load into tables
```

```
'    beta estimates
With dsResults.Tables("estimates")
    For i = 1 To UBound(objEL.beta)
        row = .NewRow
        row.Item("parameter") = objEL.beta_names(i, 1)
        row.Item("estimate") = objEL.beta(i, 1)
        row.Item("std.err") = stdErr(i)
        row.Item("ratio") = ratio(i)
        .Rows.Add(row)
    Next i
End With

' fit statistics
With dsResults.Tables("fit_stats")
    ' loglikelihood
    row = .NewRow
    row.Item("item") = "loglikelihood"
    row.Item("value") = objEL.loglik
    .Rows.Add(row)
    ' G squared
    row = .NewRow
    row.Item("item") = "G2"
    row.Item("value") = objEL.G2
    .Rows.Add(row)
    ' Chi squared
    row = .NewRow
    row.Item("item") = "X2"
    row.Item("value") = objEL.X2
    .Rows.Add(row)
    ' Deg of freedom
    row = .NewRow
    row.Item("item") = "df"
    row.Item("value") = objEL.df
    .Rows.Add(row)
End With

' covariance matrix
With dsResults.Tables("covBeta")
    For i = 1 To UBound(objEL.beta)
        For j = 1 To UBound(objEL.beta)
            row = .NewRow
            row.Item("row") = i
            row.Item("column") = j
            row.Item("value") = objEL.cov_beta(i, j)
            .Rows.Add(row)
        Next j
    Next i
End With
```

8.3.9 Finishing Up

Finally, let's add some finishing touches to our project. First, change the Text property of Form1 from `Form1` to `ELogit.NET 1.0`.

Next, add a button that allows a user to save the results to an XML file. Place a button named `pbSaveResults` and a SaveFileDialog object named `SaveFileDialog1` in the form. Add the following code to the button's event handler:

```
Dim filePath As String
Dim dsResults As DataSet

' Show dialog to get path of xml file
With SaveFileDialog1
    .Filter = "XML files (*.xml)|*.xml|" & _
        "Text files (*.txt)|*.txt|All files (*.*)|*.*"
    .DefaultExt = "*.xml"
    .ShowDialog()
    filePath = .FileName
End With

If filePath <> "" Then
    ' Write the XML file
    dsResults = DataGrid2.DataSource
    With dsResults
        .WriteXml(filePath)
    End With
End If
```

Our ELogit.NET application is finished. Compile and run the program again. After fitting a model, the results should now appear as shown in Figure 8.18. Try navigating through the tables of results in the DataGrid. Notice the tree containing hyperlinked names of the three tables (`estimates`, `fit_stats`, and `covBeta`). When you click on one of these hyperlinks, the corresponding table will appear. To return to the list of tables, click the back arrow button in the upper-right corner of the DataGrid. Alternatively, you could provide three buttons above the DataGrid to programmatically choose the table to be displayed.

More features can be added to ELogit.NET to help ensure that it is used properly. For example, we could disable the "Run Estimation" button until the model is fully specified, and we could disable the "Save Results" button until the model has been fit and the results have appeared.

8.3.10 Exercises

1. In the ELogit.NET project, add a "Save Dataset" button along with the event-handling code behind it.

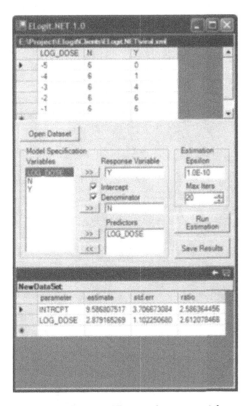

FIGURE 8.18. The finished ELogit.NET application with parameter estimates displayed in a DataGrid.

2. Create a graphical user interface for the Magic8 COM server. If you want to get fancy, incorporate a graphical representation of the ball with the text of the answer appearing in it.

3. Create a graphical user interface for a COM server that performs the ordinary chi-square test for independence in a two-way contingency table. Load the data from an XML file and write the results to an XML file.

4. ELogit.NET was designed to read data from an XML file and then load it into the ELOGIT COM server. An alternative strategy is to use the COM server's `read_datafile` and `read_namesfile` methods to first load the data into the COM server. After the data and variable names are successfully loaded, they can then be extracted from the COM server and displayed in a DataGrid. Implement this strategy and discuss the relative merits of the two alternatives. (Hint: What would happen if the user tried to edit the data in the DataGrid?)

References

Akin, E. (2003) *Object-Oriented Programming via Fortran 90/95*. New York: Cambridge University Press.

Agresti, A. (2002) *Categorical Data Analysis*, Second edition. New York: Wiley.

Box, G.E.P. and Muller, M.E. (1958) A note on the generation of random normal deviates. *Annals of Mathematical Statistics*, 29, 610–611.

Chung, H., Loken, E., and Schafer, J.L. (2004) Difficulties in drawing inferences with finite-mixture models: A simple example and a simple solution. *The American Statistician*, 58, 152–158.

Davis, B.D., Dulbecco, R. , Eisen, H.N., Ginsberg, H.S. (1989) *Microbiology*, Fourth edition. Philadelphia: Lippincott, Williams and Wilkins.

Decyk, V.K., Norton, C.D., and Szymanski, B.K. (1998) How to support inheritance and run-time polymorphism in Fortran 90. *Computer Physics Communications*, 115, 9–17.

Efron, B. and Tibshirani, R.J. (1994) *An Introduction to the Bootstrap*. New York: Chapman and Hall.

Freedman, D., Pisani, R., Purves, R (1997) *Statistics*, Third edition. New York: Norton.

Gentle, J.E. (2002) *Elements of Computational Statistics*. New York: Springer-Verlag.

Golub, G. and van Loan, C. (1996) *Matrix Computations*, Third edition. London: The Johns Hopkins University Press.

Griffiths, P. and Hill, I.D. (1985) *Applied Statistics Algorithms*. Chichester: Ellis Horwood.

Hauser, J.R. (1996) Handling floating-point exceptions in numeric programs. *ACM Transactions on Programming Languages and Systems*, 18, 139–174.

Hosmer, D.W. and Lemeshow, S. (2000) *Applied Logistic Regression*, Second edition. New York: Wiley.

Huber, P.J. (1967) The behavior of maximum likelihood estimates under non-standard conditions. In *Fifth Berkeley Symposium in Mathematical Statistics and Probability*, 221–233. Berkeley: University of California Press.

Insightful Corporation (2001) *S-PLUS 6 for Windows Programmers Guide*. Seattle, WA: Insightful Corporation.

Kahaner, D., Moler, C.B., and Nash, S. (1988) *Numerical Methods and Software*. Englewood Cliffs, NJ: Prentice-Hall.

Kennedy, W.J. and Gentle, J.E. (1980) *Statistical Computing*. New York: Marcel Dekker.

Knuth, D.E. (1971) An empirical study of FORTRAN programs. *Software—Practice and Experience*, 1, 105–133.

Knuth, D.E. (1981) *The Art of Computer Programming, Volume 2: Seminumerical Algorithms*, Second edition. Reading, MA: Addison-Wesley.

Lange, K. (1999) *Numerical Analysis for Statisticians*. New York: Springer-Verlag.

Little, R.J.A., Rubin, D.B. (2002) *Statistical Analysis with Missing Data*, Second Edition. New York: Wiley.

McCullagh, P., Nelder, J.A. (1989) *Generalized Linear Models*, Second Edition. London: Chapman and Hall.

McLachlan, G. and Peel, D. (2000) *Finite Mixture Models*. New York: Wiley.

Metcalf, M. and Reid, J. (1999) *Fortran 90/95 Explained*, Second edition. Oxford: Oxford University Press.

Myers, R.H., (1990) *Classical and Modern Regression With Applications*, Second edition. Boston: PWS.

Press, W.H., Flannery, B.P., Teukolsky, S.A., and Vetterling, W.T. (1992) *Numerical Recipes in Fortran 77: The Art of Scientific Computing*, Second edition. Cambridge: Cambridge University Press.

Press, W.H., Flannery, B.P., Teukolsky, S.A., and Vetterling, W.T. (1996) *Numerical Recipes in Fortran 90: The Art of Parallel Scientific Computing.* Cambridge: Cambridge University Press.

R Development Core Team (2004) *Writing R Extensions.* R Development Core Team.

SAS Institute Inc. (2000) *SAS Companion for the Microsoft Windows Environment, Version 8.* Cary, NC: SAS Institute Inc.

Thisted, R.A. (1988) *Elements of Statistical Computing: Numerical Computation.* London: Chapman and Hall.

Titterington, D.M., Smith, A.F.M., and Makov, U.E. (1985) *Statistical Analysis of Finite Mixture Distributions.* New York: Wiley.

White, H. (1980) Maximum likelihood estimation of misspecified models. *Econometrica*, 50, 1–25.

Index

Handbook of Computational Statistics
Concepts and Methods

J.E. Gentle, W. Härdle, and Y. Mori (Eds.)

The Handbook of Computational Statistics - Concepts and Methods is divided into 4 parts. It begins with an overview of the field of Computational Statistics, how it emerged as a seperate discipline, how it developed along the development of hard- and software, including a discussion of current active research. The second part presents several topics in the supporting field of statistical computing. Emphasis is placed on the need for fast and accurate numerical algorithms, and it discusses some of the basic methodologies for transformation, data base handling and graphics treatment. The third part focuses on statistical methodology. Special attention is given to smoothing, iterative procedures, simulation and visualization of multivariate data. Finally a set of selected applications like Bioinformatics, Medical Imaging, Finance and Network Intrusion Detection highlight the usefulness of computational statistics.

2004. 1070 p. Hardcover ISBN 3-540-40464-3

Elements of Computational Statistics

J.E. Gentle

Computationally intensive methods have become widely used both for statistical inference and for exploratory analyses of data. The methods of computational statistics involve resampling, partitioning, and multiple transformations of a dataset. They may also make use of randomly generated artificial data. Implementation of these methods often requires advanced techniques in numerical analysis, so there is a close connection between computational statistics and statistical computing. This book describes techniques used in computational statistics, and addresses some areas of application of computationally intensive methods, such as density estimation, identification of structure in data, and model building. Although methods of statistical computing are not emphasized in this book, numerical techniques for transformations, for function approximation, and for optimization are explained in the context of the statistical methods.

2002. 420 p. (Statistics and Computing) Hardcover ISBN 0-387-95489-9